Peri-urban Conflicts and Environmental Challenges: A Mediterranean Perspective

RIVER PUBLISHERS SERIES IN SOCIAL, URBAN, ECONOMIC AND ENVIRONMENTAL SUSTAINABILITY

The River Series on Social, Urban, Economic and Environmental Sustainability is a series of comprehensive academic and professional books which focus on the societal side of sustainability. The series focuses on topics ranging from theory to policy and real-life case studies and applications.

Books published in the series include research monographs, edited volumes, handbooks and textbooks. The books provide professionals, researchers, educators, and advanced students in the field with an invaluable insight into the latest research and developments.

Topics included in this series are as follows:-

- Climate Change Mitigation
- Renewable Energy Policy
- Urban sustainability
- Strategic environmental planning
- Environmental Systems Monitoring and Analysis
- Greening the World Economy
- Sustainable Development

For a list of other books in this series, visit www.riverpublishers.com

Peri-urban Conflicts and Environmental Challenges: A Mediterranean Perspective

Editors

Antonio Tomao
University of Tuscia, Italy

Matteo Clemente
University of Rome La Sapienza, Italy

LONDON AND NEW YORK

Published 2022 by River Publishers
River Publishers
Alsbjergvej 10, 9260 Gistrup, Denmark
www.riverpublishers.com

Distributed exclusively by Routledge
4 Park Square, Milton Park, Abingdon, Oxon OX14 4RN
605 Third Avenue, New York, NY 10017, USA

Peri-urban Conflicts and Environmental Challenges: A Mediterranean Perspective / by Antonio Tomao, Matteo Clemente.

Routledge is an imprint of the Taylor & Francis Group, an informa business

ISBN 978-87-7022-682-0 (Hardback)
ISBN 978-87-7022-681-3 (online)
ISBN 978-1-003-33306-7 (ebook master)

Contents

Acknowledgments

The realization of the book was supported by the LIFE project VEG-GAP "Vegetation for Urban Green Air Quality Plans" (LIFE18 PRE IT 003).

List of Figures

List of Tables

List of Contributors

Cillis, Giuseppe, *Scuola di Scienze Agrarie, Forestali, Alimentari e Ambientali, University of Basilicata, Viale dell'Ateneo Lucano, I-85100 Potenza, Italy; E-mail: giuseppe.cillis@unibas.it*

Clemente, Matteo, *Department of Architecture and Project, 'Sapienza' University of Rome, Via Flaminia 359, I-00196 Rome, Italy; E-mail: matteo.clemente@uniroma1.it*

Coluzzi, Rosa, *Institute of Methodologies for Environmental Analysis of the Italian National Research Council (IMAA–CNR), Contrada Santa Loja snc, I-85050 Tito Scalo, Italy; E-mail: rosa.coluzzi@imaa.cnr.it*

Conte, Adriano, *Italian Council for Agricultural Research and Economics (CREA), Research Centre for Forestry and Wood, Via Valle della Quistione 27, I-00166 Rome, Italy; E-mail: adriano.conte@crea.gov.it*

Ferrara, Agostino, *Scuola di Scienze Agrarie, Forestali, Alimentari e Ambientali, University of Basilicata, Viale dell'Ateneo Lucano, I-85100 Potenza, Italy; E-mail: agostino.ferrara@unibas.it*

Imbrenda, Vito, *Institute of Methodologies for Environmental Analysis of the Italian National Research Council (IMAA–CNR), Contrada Santa Loja snc, I-85050 Tito Scalo, Italy; E-mail: vito.imbrenda@imaa.cnr.it*

Lanfredi, Maria, *Institute of Methodologies for Environmental Analysis of the Italian National Research Council (IMAA–CNR), Contrada Santa Loja snc, I-85050 Tito Scalo, Italy; E-mail: maria.lanfredi@imaa.cnr.it*

Quaranta, Giovanni, *Department of Mathematics, Computer Science and Economics Department, University of Basilicata, Viale dell'Ateneo Lucano, I-85100 Potenza, Italy; E-mail: giovanni.quaranta@unibas.it*

Salvati, Luca, *Department of Methods and Models for Economics, Territory and Finance (MEMOTEF), Faculty of Economics, Sapienza University of Rome, Via del Castro Laurenziano 9, I-00161 Rome, Italy; E-mail: luca.salvati@uniroma1.it*

Salvia, Rosanna, *Department of Mathematics, Computer Science and Economics, University of Basilicata, Via dell'Ateneo Lucano 10, I-85100, Potenza, Italy; E-mail: rosanna.salvia@unibas.it*

Sorgi, Tiziano, *Italian Council for Agricultural Research and Economics (CREA), Research Centre for Forestry and Wood, Via Valle della Quistione 27, I-00166 Rome, Italy; E-mail: tiziano.sorgi@crea.gov.it*

Tomao, Antonio, *Department for Innovation in Biological, Agro-food and Forest systems (DIBAF), University of Tuscia, Via S. Camillo de Lellis, I-01100, Viterbo, Italy; E-mail: antonio.tomao@unitus*

List of Abbreviations

ADJBUI	The ratio of adjacent buildings on total buildings (2001)
AVGFSU	Average fire size by municipality (2009–2012)
CLC	Corine Land Cover
CROP	Proportion of burnt cropland on total burnt area (2009–2012)
DENPOP	Population density (2011)
DIST	Distance from Athens
DPSIR	Drivers, pressures, state, impact and response model of intervention
ELEV	Elevation (0: lowland; 1: upland municipalities)
EU	Europe
FIRDEN	Density of forest fires on total municipal area (2009–2012)
FIRSU%	Percentage of municipal area affected by fires (2009–2012)
For	Per-capita forest cover
INCOME	Average per-capita declared income (2011)
LC	LaCoast project
LCC	Land Cover Change
NSSG	National Statistical Service of Greece
PAST	Proportion of burnt pastures on total burnt area (2009–2012)
POPRAT	The ratio of present to resident population (2011)
SDG	Sustainable Development Goals
SUPMUN	Municipality surface area
UA	Urban Atlas
VARDEM	Annual population growth rate (2001–2011)
WCED	World Commission on Environment and Development
WOOD	Proportion of burnt forests on total burnt area (2009–2012)
WUI	Wildland–urban interface

1

Conflicting Landscapes: Urban Fringes and Socioeconomic Transformations in Southern Europe

Antonio Tomao[1], Rosanna Salvia[2], Matteo Clemente[3], and Luca Salvati[4]

[1]Department for Innovation in Biological, Agro-food and Forest systems (DIBAF), University of Tuscia, Via S. Camillo de Lellis, I-01100, Viterbo, Italy
[2]Department of Mathematics, Computer Science and Economics, University of Basilicata, Via dell'Ateneo Lucano 10, I-85100, Potenza, Italy
[3]Department of Architecture and Project, 'Sapienza' University of Rome, Via Flaminia 359, I-00196 Rome, Italy
[4]Department of Methods and Models for Economics, Territory and Finance (MEMOTEF), Faculty of Economics, Sapienza University of Rome, Via del Castro Laurenziano 9, I-00161 Rome, Italy
E-mail: antonio.tomao@unitus.it; rosanna.salvia@unibas.it; matteo.clemente@uniroma1.it; luca.salvati@uniroma1.it

The concept of sustainable development, meant as a state of balance between all the different components (economic, social, environmental) over time and space, represents a good contribution to complexity science, because this meaning of sustainability does not depend on a specific definition of development and also because of the possibility to define targets and mechanisms of policy and feedbacks, which involve the very same sustainability. The chance that this expectation may turn into measures of economic policies, is quite a complex topic to be investigated; nevertheless, it is linked to the relationship between the time horizon of policymakers and the impacts of non-sustainability. In such socioeconomic contexts, development may generate a negative trade-off with the environment, due to productive components of higher impact. The developing path, however, should stimulate restriction/mitigation referring to environmental degradation. We have seen how socioeconomic processes at a country level do not share a common matrix, depending on the fact that they are qualified by different context factors, such as productive, institutional and valuable, at the basis of territorial disparities (Maloutas, 2003; Harvey, 2006; Brenner et al., 2019).

Environmental mechanisms of degradation, as well as channels of transmission of social and economic impacts on ecosystems, differ on the different areas of a given Country. The usual partition into administrative regions can

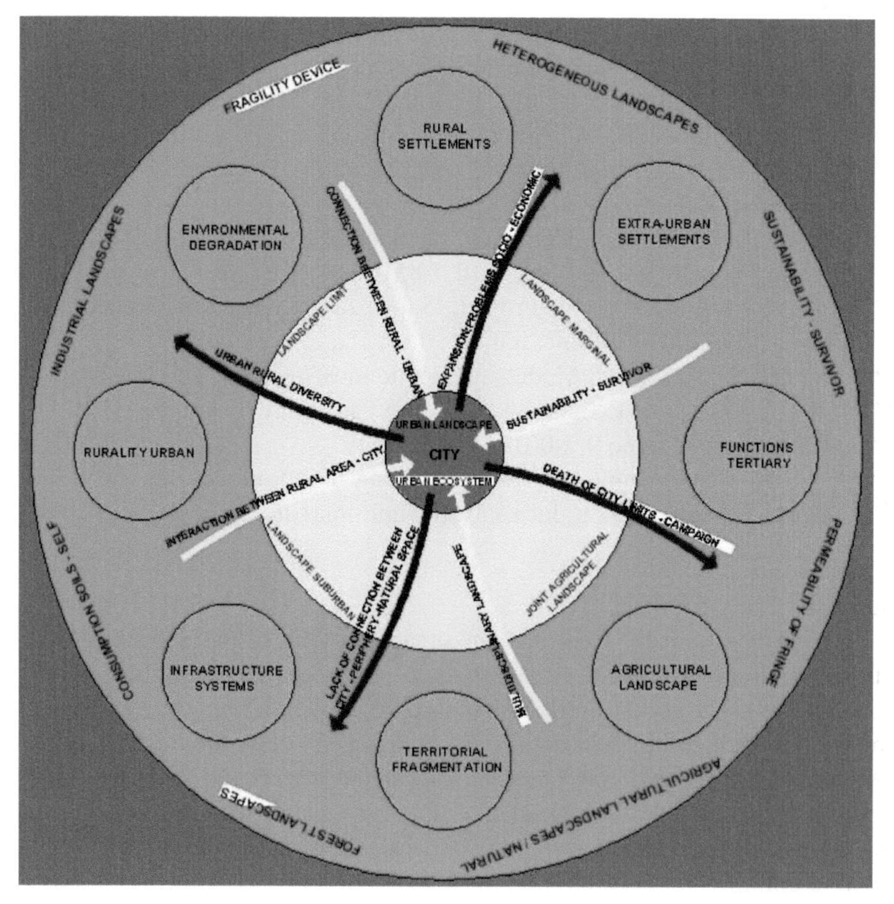

Figure 1.1 A framework interpreting socio-environmental complexity in Mediterranean areas.

Source: Authors' own elaboration.

seize, on macro-scale, significant difference within the environmental issues. Seems appropriate to point out that an articulated classification of the territory is necessary in order to orientate policies of environmental sustainability. According to Economou (1997), Giannakourou (2005), Delfanti et al. (2016) and Smiraglia et al. (2016), environmental change was interpreted as the result of a complex socio-economic system involving several interacting agents (Figure 1.1).

1.1 The 'Legacy' of Traditional Urban Forms

The notion of 'compact cities' has been clearly defined and its benefits clarified in the recent literature from a specific land-use perspective, outlining how this model is typically land-saving (Beriatos and Gospodini, 2004; Delladetsima, 2006; Arapoglou and Sayas, 2009). Although generating negative externalities on natural landscapes – especially within the boundary of urban areas (e.g. pollution, congestion, scarcity of green spaces) –, it remains a representative model of city growth with a relatively low impact on land and soils (Vaiou, 2002; Vidal et al., 2011; Zambon et al., 2018). Compact urban models are characteristic examples of multivariate relationships (e.g. in the economic, societal and environmental spheres) between the urban area and the neighboring region that are intrinsically regulated by explicit metropolitan hierarchies (Wyly, 1999). Distinctive spatial segregation in the use of land and separated economic roles for urban and rural areas are typical of complex gradients from high-density settlements to low-density peripheral villages and isolated settlements (Salvati et al., 2012a, 2012b). Cities are places of consumption and production of goods (industry and services). Rural areas are places of agricultural production and conservation (or reproduction) of natural amenities (Figure 1.2).

In some ways, discontinuous metropolises reflect a sort of 'revenge' of urban environments invading the surrounding, rural countryside through a disorganized, heterogeneous and disarticulated spillover of built-up settlements, without forming clear zoning between residential and service (commercial, industrial) districts (Recanatesi et al., 2016). The inherent transition from compact to dispersed cities results in an indistinct mix of land use classes along urban gradients, with the consequent mitigation of density divides between cores and peripheries (Marmaras, 1989; Petsimeris and Tsoulouvis, 1997; Salvati, 2014). Population shrinkage of inner cities and urban sprawl are two faces of the same medal driving a progressive reorganization of rural territories (Salvati et al., 2016). Undefined, peri-urban

Figure 1.2 The compact core of Athens, 2018.
Source: Authors' photographic archive.

landscapes consolidated, having land fragmentation as a possible keyword when interpreting its (human-derived) complexity (Prezioso, 2013). Proving the uneven disorder of land-use structures, the pristine agricultural traits of these landscapes have been (irreversibly or hardly reversibly) lost (Kourliouros, 1997). The natural features of city regions, especially forest cover, have been progressively threatened (Tomao et al., 2017). The environmental externalities of dispersed models of urban growth determine a sort of amplification of problems typical of congestion economies, merged with issues more typical of rural areas, including desertification risk, wildfires, land take (Bajocco et al., 2011, Bajocco et al., 2012; Salvati et al., 2018a, 2018b). These dynamics caused a downward spiral that is more and more difficult to assess and mitigate effectively.

The changing role of rural areas from an 'actor' (i.e. sink) of production (agriculture, livestock, forestry, natural amenities) to an 'object' at the service of the city (e.g. source of land suitable to edification) is striking

(Duvernoy et al., 2018). Are the rural areas becoming inexorably a sort of garbage can of the metropolis? Industrial and commercial areas are expelled from the city center, and re-located within the rural space (Perrin et al., 2018); residential districts reproduced themselves in the same space by obsessively standard forms, miming the compactness (and the other, well-known, problems) of the settlement types from which they conceptually derive, sprawled low-density villages grew everywhere without clear spatial planning and lacking an urban 'vision' (Emmanuel, 2014). These are all elements of a picture shot on the modern peri-urban landscape (Chorianopoulos et al., 2014). Road infrastructures (and at a less pace, railways) grew anywhere to connect the several elements that, were originally conceived as functionally distinct and spatially separated (Claval and Thompson, 1998).

Lost of the original functions of the rural area is accelerated by globalisation at all the production and spatial scales (Riley and Harvey, 2007). The 'global' citizen of the diffused city do not feel the preservation of the agricultural base close to the city as a crucial planning issue, as the primary goods are provided to the increasing urban markets from very far lands, better equipped for large-scale production (Juntti and Wilson, 2005). The same citizen could even evaluate as superfluous to protect the remaining (not sealed) rural territory which produces natural amenities and tourism leisure, as increasing human mobility stimulates tourism fluxes concentrating well outside the boundary of the urban region and big pipelines bring to the city the (supposedly unlimited) quantity of water produced far from its surroundings (Imeson, 2012).

1.2 A Shift Towards the 'Sprawling' City

The rural space is becoming, in the Mediterranean city-region, a mirror of the urban landscape, where not only rich people built-up their villas (in a typical 'lock living' framework) but where also bourgeoises and even poor people escape the congestion of the city centre and the compact periphery searching for 'deconcentration' amenities (Maloutas, 2007). Those second homes are inhabitated not only during summer holidays but also during the free time all over the year and likely more (Leontidou, 1996). By escaping the urban congestion, however, humans create another type of congestion, which is the loss of the (hierarchical) order of rural landscapes close to the traditional, Mediterranean compact city (Kourliouros, 2003).

The dispersed second-home settlements (with all ranges of construction possibilities, from the poor 'villages' to the rich villas) are only one of the proofs for a 'landscape congestion': they not only dominate the 'littoralised' coastal landscape, but also invade the internal lowlands, originally inhabited by crops, fruit trees, and grapevines and even the sloping uplands, the original kingdom of olives and forests (Katsibokis, 2013). A proof that will anticipate the uneven environmental (and socio-economic) problems starting from weak management of the dispersed urban settlements within a landscape that has lost its rural traits and urges dedicated policy strategies well beyond the standard urban planning tools (Pumain, 2003).

The present contribution discusses recent findings in environmental issues dealing with desertification risk and regional disparities in the Mediterranean basin. By focusing on key socioeconomic factors (population growth, urban sprawl, coastalization, crop intensification, land abandonment) underlying land and soil degradation, this contribution highlights the intimate link between socioeconomic processes, rural poverty and territorial disparities based on complex dynamics of demographic and economic factors (Gospodini, 2001; Petsimeris, 2002; Serra et al., 2014). The increasing complexity in the spatial distribution of land vulnerable to degradation has been also pointed out with special reference to post-war Mediterranean Europe as the result of non-linear biophysical and socioeconomic dynamics (Salvati and Zitti, 2009). The lack of multi-target and multiscale policies approaching together land degradation and territorial disparities is finally discussed as an original contribution to the study of Mediterranean environments (Ciommi et al., 2018).

Effective measures against land degradation, according to the concept of sustainable development, should go further than a restricted sectorial vision, trying to pursuit a multi-target and multi-scale approach (Linke and O'Loughlin, 2015). This is of great importance when different degradation processes occur in synergy at the local level, with interactions and feedbacks that cannot be managed only by sectorial environmental measures (Kelly et al., 2015). The 'local' dimension finally remains the basis for any further strategy of intervention (no matter if on a national or over national scale) and should involve specific actions to mitigate territorial disparities (Skayannis, 2013). To be pointed out is the need to consider the increased territorial disparities, concerning the distribution of natural resources as a matter of socio-environmental equity, with further implications into the economic field (e.g. Salvati et al., 2019). In the lines above we tried to point out how the inherent 'complexification' of land degradation geography in the Mediterranean basin and the increasing interactions between different gradients, which determine its space and time distribution, could lead to increase environmental disparities and consolidate territories which are traditionally polarized (Di Feliciantonio et al., 2015). This phenomenon is to be integrated with the general overview of environmental policies on a national and European scale.

1.3 Urban Challenges and Peri-Urban Conflicts

The interactions between land degradation and human pressure (differently declined as described in this contribution, such as poverty, territorial disparities, urban-rural polarization, the transformation of productive structures), may stimulate general and local measures of environmental policies, in order to face the new issues proposed by territorial development, that is to say, the rebalancing of the population density gradients and the territory development through a polycentric vision (Coccossis et al., 2005). It is very important to consider the restriction of causes that have generated inner migrations, in order to affect phenomena such as coastalization and urban sprawl in Mediterranean Europe (Maloutas, 2012; Cuadrado-Ciuraneta et al., 2017; Di Feliciantonio et al., 2018). Scientific research assumes great importance, allowing the chance to point out policies to activate in different contexts, according to the global institutional context, all the possible strategies to be pursued and the overall degree of knowledge of phenomena and their interactions with biophysical and socioeconomic dimensions (Vaiou, 1997).

Chapter 2 investigates trends in population growth and urbanization characteristic of some specific examples in the Mediterranean region. 'Coastalisation', a general urban development pattern of the region, is analyzed and discussed, taking advantage of various data sources, which include statistical data, maps, photographs and literature review. A concise discussion was deserved to demonstrate the connection between over-urbanisation and the growing vulnerability of cities over environmental crises (Kaika, 2012). Specific analysis of the observed consequences of urban sprawl was finally presented.

In the subsequent chapters, a specific focus was deserved to a paradigmatic metropolis in the Mediterranean region, by delineating urbanization and suburbanization trends in Attica, a socially divided and economically polarized metropolitan area including Athens, the capital city of Greece (Kandylis et al., 2012). Attica was partitioned in two spatial domains (the Greater Athens district, including Athens and Piraeus Figure 1.3, and the rest of the metropolitan region administered by more than 50 municipalities).

Figure 1.3 A summary geography of the Athens' metropolitan region, representing the vast majority of the administrative region of Attica, Central Greece.
Source: Authors' own elaboration.

Two distinct growth waves (with a breakpoint in the early 1980s) were identified, distinguishing 'fast' from 'slow' changes in the factors contributing to Athens' transition from industry to advanced services (Gospodini, 2006; Gospodini, 2009; Souliotis, 2013).

Up to the early 1980s, Greater Athens experienced population increase and settlement densification; the rest of Attica underwent moderate suburbanization with expanding urban functions (Roubien, 2017; Nickayin et al., 2020). Since the late 1980s, greater Athens experienced stable population and economic consolidation; the rest of Attica underwent residential sprawl and concentration of activities in the most accessible suburbs (Salvati, 2016). This spatial pattern reflects the specificity of recent Mediterranean urbanization in respect to the development paths prevailing in western and northern European cities (Panori et al., 2019). Empirical evidence highlights the importance of multi-scale analysis investigating growth and change of informal and scattered cities in both developed and emerging countries (Kazemzadeh-Zow et al., 2017).

Chapter 3 analyzes the spatial distribution of wildfires recorded during recent years in a spatial partition of nearly 60 municipalities in Attica. Assuming high fire risk and significant human pressure driven by urban expansion, the hypothesis that a defined fire profile (in terms of density, severity and land-use selectivity) at the local scale was associated with a specific set of socioeconomic and territorial variables, was tested explicitly using six fires' indicators and eight contextual indicators under a multivariate analysis framework. The analysis identified two main dimensions for both forest fires (dimension and selectivity) and the socioeconomic context (demographic variables associated with the urban-rural gradient and average income). Fire density and forest/pastures burnt areas were not correlated to any socioeconomic variable. At the same time, the average declared income and elevation of each municipality did not correlate to any fires' variable. On the contrary, the average fire size, the percentage of burnt area per municipality and the proportion of cropland affected by fires correlated positively with the distance from the inner city and the total surface area of each municipality and negatively with the proportion of compact settlements, population density and growth. These results confirm the impact of the urban-to-rural gradient in the spatial distribution of forest fires in Attica (Pili et al., 2017) while pointing out the negligible influence of variables such as the socioeconomic status of the resident population.

Chapter 4 illustrates a comprehensive analysis of an indicator of per-capita forest area in the districts of Attica, a Mediterranean urban region

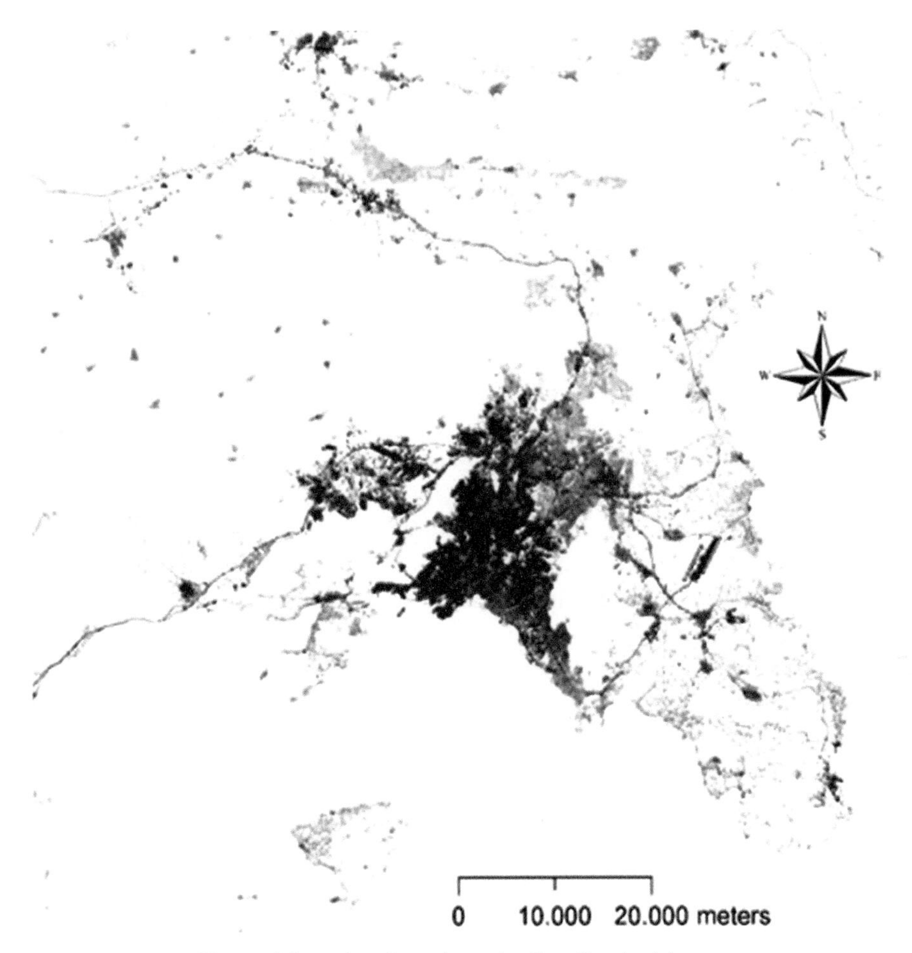

Figure 1.4 Urban footprint and soil sealing in Athens.
Source: Authors' own elaboration on European Environment Agency data.

where urban expansion was not regulated by integrated planning and successive waves of spontaneous growth occurred (Figure 1.4).

It was demonstrated how forest land per inhabitant showed a progressive decrease between 1960 and 2010. On the one hand, urban areas surrounding the centers of Athens and Piraeus experienced low levels of forest availability per inhabitant since 1960. In addition high forest cover per capita progressively declined (1980, 1990) along fringe areas surrounding Athens and in 'sprawl' districts (such as Messoghia). In the last few years, some municipalities showed an increase in forest cover, mainly due to land abandonment and

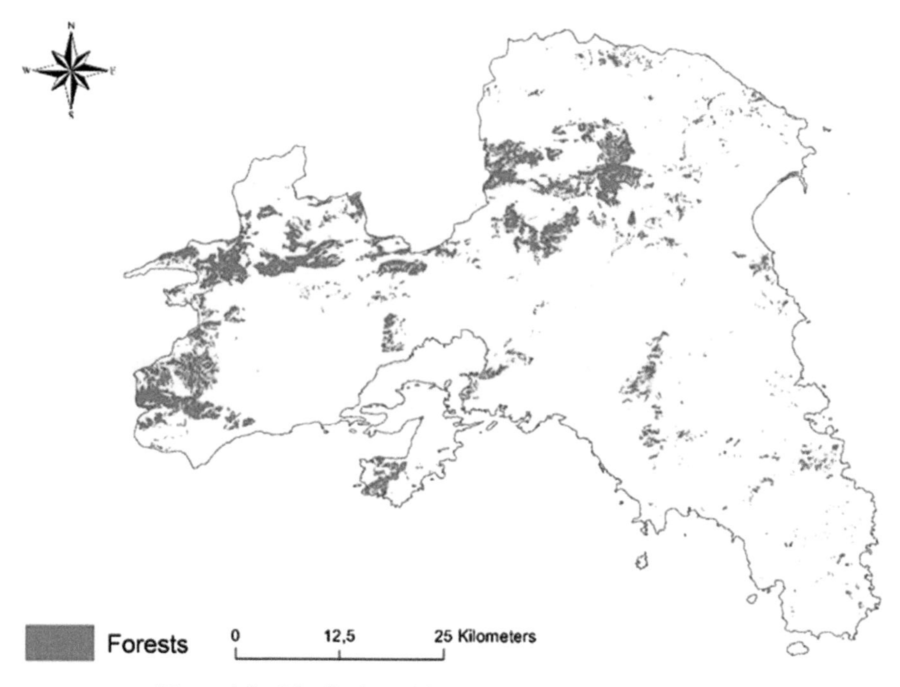

Figure 1.5 Distribution of forest cover in the region of Attica.
Source: Authors' own elaboration on Urban Atlas data.

forest recolonization (Figure 1.5). Findings from this study suggest that more effective protection of forests and agricultural land is required, especially for rural and remote districts (Biasi et al., 2015).

Long-term urban expansion in Southern European metropolises has been finally investigated in Chapter 5 by interpreting the overall vision and the practical objectives of planning.The landscape is a complex system of interrelated biophysical and anthropogenic elements that changes rapidly according to external pressures. This system definitely needs permanent monitoring systems to inform effective conservation policies. Landscape analysis benefits of integrated approaches encompassing geography, environmental science and information systems to evaluate structure, form, diversity and dynamics of different land-use types (Figure 1.6). While nowadays monitoring techniques for land-use changes detection are common and well defined, the evaluation of specific landscape elements over time is a research issue that deserves further efforts in both the theoretical and practical perspectives. The Mediterranean rural landscape is one of the richest in terms of cultural and natural

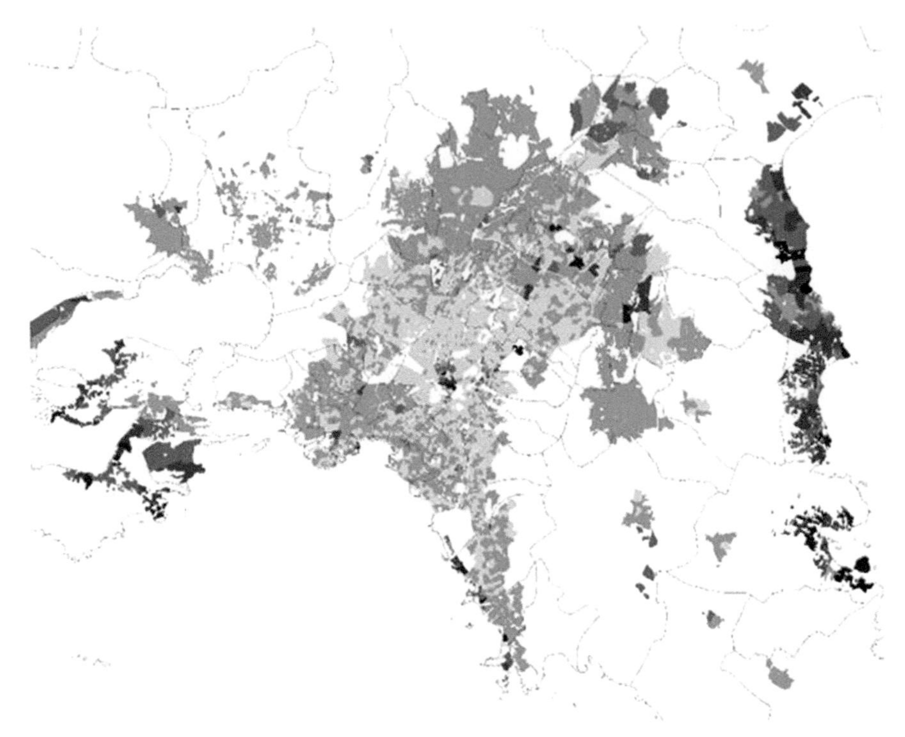

Figure 1.6 The ratio between resident and present population at the national census, 2001, Athens' census tracks (darkest tonalities in Messoghia, Eastern Attica and Salamina island, Western Attica, indicate present population above resident population in settlement neighborhoods).

Source: Authors' elaboration on Census track data provided by Hellenic Statistical Authority.

biodiversity in the world since it is characterized by specific structures and multifaceted functions. This landscape type, however, is threatened by drastic land-use changes in the area mainly driven by urbanization, infrastructural development and cropland abandonment.

Diverging town planning orientations overtime in Athens have reflected the slow evolution towards a less compact and mono-centric spatial asset (Di Feliciantonio et al., 2018), typical of several Mediterranean cities. By indirectly sustaining (or tolerating) informal housing, sequential master plans in Athens, at least until the late 1980s, have progressively incorporated discontinuous or isolated settlements in the consolidated urban fabric, creating a mixed city model suspended in between compactness and dispersion (Carlucci et al., 2018). The study finally argues how this model is a result of multiple social forces competing in the urban arena (Ciommi et al., 2017).

Earlier findings clearly documented how dispersed urbanization during the last half-century has transformed metropolitan regions into well-connected, low-density residential areas (Ciommi et al., 2019). However, this kind of urbanization has changed irreversibly the traditional rural landscape around cities, leading to a new definition of rurality (Salvati and Serra, 2016).

1.4 Concluding Remarks

Urban expansion and the preservation of fringe landscapes are clearly interconnected issues. This book discusses the relationship between landscape and peri-urban agriculture and the possible implications of sustainable land management for fringe land quality, proposing a framework to evaluate the latent nexus between agro-forest systems and human settlements in Southern Europe. Eco-sustainable planning integrated with multifaceted policy actions (e.g. social, economic, cultural and political dimensions) is a relevant approach to reinforce the sustainability of fringe landscapes. Permanent assessment of these factors allows for the implementation of different development scenarios. The present study definitely contributes to systemic and multi-scale approaches informing environmental policies, with the aim of achieving integrated management of peri-urban rural landscapes.

References

Arapoglou, V. P., and Sayas, J. (2009). New facets of urban segregation in southern Europe: Gender, migration and social class change in Athens. *European Urban and Regional Studies*, 16(4), pp. 345–362.

Bajocco, S., De Angelis, A., Salvati, L. (2012). A satellite-based green index as a proxy for vegetation cover quality in a Mediterranean region. *Ecological Indicators*, 23, pp. 578–587.

Bajocco, S., Salvati, L., and Ricotta, C. (2011). Land degradation versus fire: A spiral process? *Progress in Physical Geography*, 35(1), pp. 3–18.

Beriatos, E., and Gospodini, A. (2004). "Glocalising" urban landscapes: Athens and the 2004 Olympics. *Cities*, 21(3), pp. 187–202.

Biasi, R., Colantoni, A., Ferrara, C., Ranalli, F., and Salvati, L. (2015). In-between sprawl and fires: Long-term forest expansion and settlement dynamics at the wildland–urban interface in Rome, Italy. *International Journal of Sustainable Development & World Ecology*, 22(6), pp. 467–475.

Brenner, N., Peck, J., and Theodore, N. (2010). Variegated neoliberalization: geographies, modalities, pathways. *Global Networks*, 10(2), pp. 182–222.

Carlucci, M., Chelli, F.M., Salvati, L. 2018. Toward a new cycle: Short-term population dynamics, gentrification, and re-urbanization of Milan (Italy). *Sustainability* (Switzerland) 10(9), 3014.

Chorianopoulos, I., Tsilimigkas, G., Koukoulas, S., and Balatsos, T. (2014). The shift to competitiveness and a new phase of sprawl in the Mediterranean city: Enterprises guiding growth in Messoghia–Athens. *Cities*, 39, pp. 133–143.

Ciommi, M., Chelli, F.M., Carlucci, M., Salvati, L. 2018. Urban growth and demographic dynamics in southern Europe: Toward a new statistical approach to regional science. *Sustainability* (Switzerland), 10(8), 2765.

Ciommi, M., Chelli, F.M., Salvati, L. 2019. Integrating parametric and non-parametric multivariate analysis of urban growth and commuting patterns in a European metropolitan area. *Quality and Quantity* 53(2), pp. 957–979.

Ciommi, M., Gentili, A., Ermini, B., Chelli, F.M., Gallegati, M. 2017. Have Your Cake and Eat it Too: The Well-Being of the Italians (1861–2011). *Social Indicators Research* 134(2), pp. 473–509.

Claval, P., and Thompson, I. B. (1998). An introduction to regional geography (pp. 282–283). Oxford: Blackwell.

Coccossis, H., Economou, D., and Petrakos, G. (2005). The ESDP relevance to a distant partner: Greece. *European Planning Studies*, 13(2), pp. 253–264.

Cuadrado-Ciuraneta, S., Durà-Guimerà, A., and Salvati, L. (2017). Not only tourism: Unravelling suburbanization, second-home expansion and 'rural' sprawl in Catalonia, Spain. *Urban Geography*, 38(1), pp. 66–89.

Delfanti, L., Colantoni, A., Recanatesi, F., Bencardino, M., Sateriano, A., Zambon, I., and Salvati, L. (2016). Solar plants, environmental degradation and local socioeconomic contexts: A case study in a Mediterranean country. *Environmental Impact Assessment Review*, 61, pp. 88–93.

Delladetsima, P. M. (2006). The emerging property development pattern in Greece and its impact on spatial development. *European Urban and Regional Studies*, 13(3), pp. 245–278.

Di Feliciantonio, C., Salvati, L. (2015). 'Southern' Alternatives of Urban Diffusion: Investigating Settlement Characteristics and Socio-Economic Patterns in Three Mediterranean Regions. Tijdschrift voor economische en sociale geografie, 106(4), pp. 453–470.

Di Feliciantonio, C., Salvati, L., Sarantakou, E., and Rontos, K. (2018). Class diversification, economic growth and urban sprawl: evidences from a pre-crisis European city. *Quality & Quantity*, 52(4), pp. 1501–1522.

Duvernoy, I., Zambon, I., Sateriano, A., and Salvati, L. (2018). Pictures from the other side of the fringe: Urban growth and peri-urban agriculture in a post-industrial city (Toulouse, France). *Journal of Rural Studies*, 57, pp. 25–35.

Economou, D. (1997). The planning system and rural land use control in Greece: A European perspective. *European Planning Studies*, 5(4), pp. 461–476.

Emmanuel, D. (2014). The Greek system of home ownership and the post-2008 crisis in Athens. Région et Développement, 39, pp. 167–182.

Giannakourou, G. (2005). Transforming spatial planning policy in Mediterranean countries: Europeanization and domestic change. *European Planning Studies*, 13(2), pp. 319–331.

Gospodini, A. (2001). Urban design, urban space morphology, urban tourism: an emerging new paradigm concerning their relationship. *European Planning Studies*, 9(7), pp. 925–934.

Gospodini, A. (2006). Portraying, classifying and understanding the emerging landscapes in the post-industrial city. *Cities*, 23(5), pp. 311–330.

Gospodini, A. (2009). Post-industrial trajectories of Mediterranean European cities: the case of post-Olympics Athens. *Urban Studies*, 46(5–6), pp. 1157–1186.

Harvey, D. (2006). The geographies of critical geography. *Transactions of the Institute of British Geographers*, pp. 409–412.

Imeson, A. (2012). Desertification, land degradation and sustainability. *John Wiley & Sons*.

Juntti, M., and Wilson, G. A. (2005). Conceptualizing desertification in Southern Europe: stakeholder interpretations and multiple policy agendas. *European Environment*, 15(4), pp. 228–249.

Kaika, M. (2012). The economic crisis seen from the everyday: Europe's nouveau poor and the global affective implications of a 'local' debt crisis. City, 16(4), pp. 422–430.

Kandylis, G., Maloutas, T., and Sayas, J. (2012). Immigration, inequality and diversity: socio-ethnic hierarchy and spatial organization in Athens, Greece. *European Urban and Regional Studies*, 19(3), pp. 267–286.

Katsibokis, G. (2013). Ktiriothiki: The Architectural Heritage of Athens, 1830–1950. *Journal of Modern Greek Studies*, 31(1), pp. 133–149.

Kazemzadeh-Zow, A., Zanganeh Shahraki, S., Salvati, L., and Samani, N. N. (2017). A spatial zoning approach to calibrate and validate urban growth models. *International Journal of Geographical Information Science*, 31(4), pp. 763–782.

Kelly, C., Ferrara, A., Wilson, G. A., Ripullone, F., Nolè, A., Harmer, N., and Salvati, L. (2015). Community resilience and land degradation in forest and shrubland socio-ecological systems: Evidence from Gorgoglione, Basilicata, Italy. *Land Use Policy*, 46, pp. 11–20.

Kourliouros, E. (1997). Planning industrial location in Greater Athens: The interaction between deindustrialization and anti-industrialism during the 1980s. *European Planning Studies*, 5(4), pp. 435–460.

Kourliouros, E. (2003). Reflections on the economic-noneconomic debate: A radical geographical perspective from the european south. *Antipode*, 35(4), pp. 781–799.

Leontidou, L. (1996). Alternatives to modernism in (southern) urban theory: Exploring in-between spaces. *International Journal of Urban and Regional Research*, 20(2), pp. 178–195.

Linke, A. M., and O'Loughlin, J. (2015). Spatial analysis. The Wiley Blackwell companion to political geography, 189.

Maloutas, T. (2003). Promoting social sustainability The case of Athens. City, 7(2), pp. 167–181.

Maloutas, T. (2007). Segregation, social polarization and immigration in Athens during the 1990s: theoretical expectations and contextual difference. *International Journal of Urban and Regional Research*, 31(4), pp. 733–758.

Maloutas, T. (2012). Contextual diversity in gentrification research. *Critical Sociology*, 38(1), pp. 33–48.

Marmaras, E. (1989). The privately-built multi-storey apartment building: The case of inter-war Athens. *Planning Perspective*, 4(1), pp. 45–78.

Nickayin, S. S., Tomao, A., Quaranta, G., Salvati, L., and Gimenez Morera, A. (2020). Going toward Resilience? Town Planning, Peri-Urban Landscapes, and the Expansion of Athens, Greece. *Sustainability*, 12(24), pp. 10471.

Panori, A., Psycharis, Y., and Ballas, D. (2019). Spatial segregation and migration in the city of Athens: Investigating the evolution of urban socio-spatial immigrant structures. *Population, Space and Place*, 25(5), e2209.

Perrin, C., Nougarèdes, B., Sini, L., Branduini, P., and Salvati, L. (2018). Governance changes in peri-urban farmland protection following decentralisation: A comparison between Montpellier (France) and Rome (Italy). *Land Use Policy*, 70, pp. 535–546.

Petsimeris, P. (2002). Population deconcentration in Italy, Spain and Greece: A first comparison. *Ekistics*, pp. 163–172.

Petsimeris, P., and Tsoulouvis, L. (1997). Current trends and prospects of future change in the Greek urban system. Rivista geografica italiana, 104(4), pp. 421–443.

Pili, S., Grigoriadis, E., Carlucci, M., Clemente, M., and Salvati, L. (2017). Towards sustainable growth? A multi-criteria assessment of (changing) urban forms. *Ecological Indicators*, 76, pp. 71–80.

Prezioso, M. (2013). Geographical and territorial vision facing the crisis. *Journal of Global Policy and Governance*, 2(1), pp. 27–44.

Pumain, D. (2003). Une approche de la complexité en géographie. Géocarrefour, 78(1), pp. 25–31.

Recanatesi, F., Clemente, M., Grigoriadis, E., Ranalli, F., Zitti, M., and Salvati, L. (2016). A fifty-year sustainability assessment of Italian agro-forest districts. *Sustainability*, 8(1), pp. 32.

Riley, M., and Harvey, D. (2007). Talking geography: on oral history and the practice of geography.

Roubien, D. (2017). Creating modern Athens: a capital between east and west. Taylor & Francis, London.

Salvati, L. (2014). Towards a Polycentric Region? The Socio-economic Trajectory of Rome, an 'Eternally Mediterranean' City. Tijdschrift voor economische en sociale geografie, 105(3), pp. 268–284.

Salvati, L. (2016). The Dark Side of the Crisis: Disparities in per Capita income (2000–12) and the Urban-Rural Gradient in Greece. Tijdschrift voor economische en sociale geografie, 107(5), pp. 628–641.

Salvati, L., Carlucci, M., Grigoriadis, E., Chelli, F.M. 2018b. Uneven dispersion or adaptive polycentrism? Urban expansion, population dynamics and employment growth in an 'ordinary' city. *Review of Regional Research* 38(1), pp. 1–25.

Salvati, L., Ciommi, M.T., Serra, P., Chelli, F.M. 2019. Exploring the spatial structure of housing prices under economic expansion and stagnation: The role of socio-demographic factors in metropolitan Rome, Italy. *Land Use Policy* 81, pp. 143–152.

Salvati, L., Ferrara, A., Chelli, F. 2018a. Long-term growth and metropolitan spatial structures: an analysis of factors influencing urban patch size under different economic cycles. Geografisk Tidsskrift/Danish *Journal of Geography* 118(1), pp. 56–71.

Salvati, L., Gemmiti, R., Perini, L. (2012b). Land degradation in Mediterranean urban areas: an unexplored link with planning? Area, 44(3), pp. 317–325.

Salvati L., Perini L., Sabbi A., Bajocco S. (2012a). Climate Aridity and Land Use Changes: A Regional-Scale Analysis. *Geographical Research* 50(2), pp. 193–203.

Salvati, L., Sateriano, A., Grigoriadis, E. (2016). Crisis and the city: profiling urban growth under economic expansion and stagnation. *Letters in Spatial and Resource Sciences*, 9(3), pp. 329–342.

Salvati, L., and Serra, P. (2016). Estimating rapidity of change in complex urban systems: A multidimensional, local-scale approach. *Geographical Analysis*, 48(2), pp. 132–156.

Serra, P., Vera, A., Tulla, A. F., and Salvati, L. (2014). Beyond urban–rural dichotomy: Exploring socioeconomic and land-use processes of change in Spain (1991–2011). *Applied Geography*, 55, pp. 71–81.

Salvati, L., and Zitti, M. (2009). The environmental 'risky' region: identifying land degradation processes through integration of socio-economic and ecological indicators in a multivariate regionalization model. *Environmental Management*, 44(5), pp. 888.

Skayannis, P. (2013). The (master) plans of Athens and the challenges of its re-planning in the context of crisis. ArchNet-IJAR: *International Journal of Architectural Research*, 7(2), pp. 192.

Smiraglia, D., Ceccarelli, T., Bajocco, S., Salvati, L., and Perini, L. (2016). Linking trajectories of land change, land degradation processes and ecosystem services. *Environmental Research*, 147, pp. 590–600.

Souliotis, N. (2013). Cultural economy, sovereign debt crisis and the importance of local contexts: The case of Athens. Cities, 33, pp. 61–68.

Tomao, A., Quatrini, V., Corona, P., Ferrara, A., Lafortezza, R., and Salvati, L. (2017). Resilient landscapes in Mediterranean urban areas: Understanding factors influencing forest trends. *Environmental Research*, 156, pp. 1–9.

Vaiou, D. (1997). Facets of spatial development and planning in Greece. *European Planning Studies* 5, pp. 431–433.

Vaiou, D. (2002). Milestones in the urban history of Athens. Treballs de la Societat Catalana de Geografia, pp. 209–226.

Vidal, M., Domene, E., and Sauri, D. (2011). Changing geographies of water-related consumption: residential swimming pools in suburban Barcelona. *Area*, 43(1), pp. 67–75.

Wyly, E. K. (1999). Continuity and change in the restless urban landscape. *Economic Geography*, 75(4), pp. 309–338.

Zambon, I., Benedetti, A., Ferrara, C., Salvati, L. (2018). Soil matters? A multivariate analysis of socioeconomic constraints to urban expansion in Mediterranean Europe. *Ecological Economics*, 146, pp. 173–183.

2

Densifying, Decompacting, Rethinking Cities: A Mediterranean Debate

Antonio Tomao[1], Tiziano Sorgi[2], Matteo Clemente[3], and Luca Salvati[4]

[1]Department for Innovation in Biological, Agro-food and Forest systems (DIBAF), University of Tuscia, Via S. Camillo de Lellis, I-01100, Viterbo, Italy
[2]Italian Council for Agricultural Research and Economics (CREA), Research Centre for Forestry and Wood, Via Valle della Quistione 27, I-00166 Rome, Italy
[3]Department of Architecture and Project, 'Sapienza' University of Rome, Via Flaminia 359, I-00196 Rome, Italy
[4]Department of Methods and Models for Economics, Territory and Finance (MEMOTEF), Faculty of Economics, Sapienza University of Rome, Via del Castro Laurenziano 9, I-00161 Rome, Italy
E-mail: antonio.tomao@unitus.it; tiziano.sorgi@crea.gov.it; matteo.clemente@uniroma1.it; luca.salvati@uniroma1.it

Abstract

In the chapter, trends regarding population growth and urbanization rates of the Mediterranean region will be analysed. Successively, taking advantage of the analysis of urban dynamics of a paradigmatic example of a Mediterranean city (Athens), some words will be spent in demonstrating the connection between over-urbanization and the growing vulnerability of cities towards natural disasters. Finally, an analysis on the observed consequences of urban sprawl—and more specifically of 'coastalisation', the general urban development pattern of the region—will be presented.

Keywords: Mediterranean city, Urban form, Land consumption, Exurban development, Planning.

2.1 Introduction

Coastal territories represent an important resource for Europe since they contribute to the local economy (e.g., tourism), human well-being (recreation opportunities) and harbor a large number of natural well-preserved habitats (Gasparella et al., 2017; Tomao et al., 2018). The growing importance of coastal territories has led to substantiating positive population dynamics of this terrestrial surface. Coasts of EU Member States are also the most densely populated zones in Europe and particularly of Mediterranean countries (UNEP/MAP and Plan Bleu, 2020). According to various studies (e.g., Kurt, 2016; Valev, 2009), coasts include around 12% of Europe's municipalities and more than a third of its total population. Among the countries of the European Union, 40% of the territory and 40.8% of the population lived there in 2011 (Eurostat regional yearbook 2012). This phenomenon of concentration of population and economic activities in coastal areas with consequent urbanization of these territories has been defined as 'coastalisation' (Bell et al., 2013; Kasanko et al., 2006; Leontidou, 1990; Morelli and Salvati, 2010; Salvati and Forino, 2014; Sreeja et al., 2016; Mikhaylov et al., 2018).

Such shift towards the sea, also named as 'coastal sprawl' has been described for numerous Mediterranean cities (Salvati, 2014; Salvati et al., 2016), where urban pressure largely increased during the last 50 years when built areas doubled or more than doubled within one kilometre from the sea (UNEP/MAP and Plan Bleu, 2020). Moreover, urbanization rates are expected to further increase throughout the entire region in the next decades (Salvati and Serra, 2016).

These phenomena have been associated with a relevant change in urban expansion models. In the last century, several Mediterranean cities have changed their shape from the classic and traditional 'compact' form in various 'dispersed' forms, characterised by a relevant and very fast sprawl around the urban area (Schneider and Woodcock, 2008). After the Second World War, Mediterranean cities have expanded as a result of consecutive waves of compact and dispersed urbanization that formed highly polarized metropolitan regions. Actually, a 'deconcentration' phase can be recognised in Mediterranean Europe, as urban urbanization is expanding at much faster rates than population growth. Examples of this trend can be found in Lisbon (Mascarenhas et al., 2019), Barcelona (Zambon et al., 2017; Pagliarin, 2018), Rome (Di Feliciantonio and Salvati, 2015), Athens (Pili et al., 2017) and many other cities. The diffusion of these sparse, low-density settlements has witnessed over-densification of central districts but has also caused the homogenisation of urban environments and the standardisation of the landscape (Salvati et al., 2016).

The reasons for such urbanization trend are manifold and vary between cities, regions and countries in Europe. Driving forces range from economic (e.g. globalization, economic development) to societal (demographic dynamics) and governance (deregulation and informality) factors and are therefore dependent on the political, social and economic conditions in each city (Christiansen and Loftsgarden, 2011). The study of these 'territorial symptoms' of undefined evolution can provide useful information to policy-makers to adopt the correct measures with the need to preserve the natural and human environment.

With these premises, the objective of this research is to analyze the recent dynamics of urban transformation of the Mediterranean area and understand the reasons that have led to the consequent urban sprawl. For the purpose of this chapter, two issues currently debated within the 'urban geography' and 'landscape' disciplines are examined by considering the coastal metropolis of Athens as an exemplificative case study:

1. Rural-urban migration and socioeconomic disparities;
2. Coastal urbanization and environmental changes threatening natural capital and cultural heritage.

The loss of areas originally devoted to arable lands, pastures, vineyards, and annual crops is only the most visible effect in metropolitan regions. A more subtle phenomenon transforming, fragmenting, simplifying, and deteriorating the rural landscape is in progress. While in the past, urban growth occurred primarily on agricultural and semi-natural areas at the urban fringe, now sprawl invaded larger areas more and more distant from the city centre (Bruegmann, 2005; Tomao et al., 2017).

2.2 Mediterranean Urbanization

The urbanization of the Mediterranean region of Europe has a long and complex history which has led to distinctive and original urban forms. The scientific literature has largely debated about how to define an archetype of 'Mediterranean city' based on the similarities in socioeconomic development patterns and urban morphology (e.g. Leontidou, 1996; Salvati and Gargiulo Morelli, 2014). The economic and social structures of the cities of the northern Mediterranean are justified by a peculiar history of interactions between demographic, cultural and political factors that featured unique landscapes

(King et al., 1997). However, despite Southern Europe has been reported as generally homogeneous in terms of urban structure (e.g. Leontidou, 1996; Colantoni et al., 2016), several scholars have highlighted how both dense and dispersed urban models are present in the Northern part of the Mediterrnaean basin (Couch et al., 2007). Leontidou (1990) has described the Mediterranean city with a 'hyper-compact' form characterized by dense settlements with poor infrastructural networks and public services, and a 'popular land control' manifested by informal settlements. In particular, the latter is rather common expansion processes across the decades of the 20th century in several Mediterranean cities (King et al., 1997). Such a spontaneous urban expansion, coupled with planning deregulation, has produced sociallyfragmented landscapes with a spatial pattern difficult to label as 'polycentric' (Maloutas, 2007; De Muro et al., 2011).

As an additional outcome of deregulated urban expansion is social stratification and economic polarization (Malheiros, 2002; Arapoglou et al., 2009). The socio-spatial pattern based on the 'inverse Burgess model' has lasted up to the 1980s in many different Mediterranean contexts (Leontidou, 1990). Furthermore, it was associated not only to economic informality and planning deregulation, but also to family-oriented welfare regimes and (partly) ineffective housing policies (Arbaci, 2007; Arbaci and Malheiros, 2010). Among the push- and pull-factors that conditioned urbanization in the Mediterranean context, population growth has been a key driver in the rapid expansion of large cities (Figure 2.1). The demographic data of Mediterranean countries indicate two different patterns (Figure 2.2). Between 1850 and 2000, the population of the Southern European EU-member countries doubled. During the same period, the rest of the Mediterranean countries showed a 9-fold increase in population (Table 2.1). In 2050, a decrease in population is expected in the five Southern European countries, while major increases in North Africa (+97 million) and in the Eastern Mediterranean (+84.3 million) are projected. Furthermore, in the 12 Middle East and North African countries, more people will be added until 2050 than presently live in the five Southern EU countries (177.3 million).

If compared to the rest of the world, where half of the population lives in urban areas, two-thirds of inhabitants in the countries bordering the Mediterranean already live in towns (United Nations, 2019). In the Mediterranean coastal region, the population went from 285 million in 1970 to 427 million in 2000 and will probably overcome 400 million in 2050. This represents a real challenge in the Mediterranean coastal region, especially in the light of the Agenda for Sustainable Development (SDG Target 11.1 expressing the need

Figure 2.1 Compact settlements in Kallithea, central Athens prefecture.
Source: Authors' photographic archive.

to 'make cities and human settlements inclusive, safe, resilient and sustainable'). Moreover, Mediterranean coasts are threatened by changing climate conditions (UNEP/MAP and Plan Bleu, 2020) and the potential impacts of this phenomenon are further exacerbated by the process of coastalization due to a combination of specific social, economic and environmental characteristics such as rapid rural to urban migration and mass tourism. Indeed, in the last years of the 20th century, the increase of international tourism has invaded the shores of the Mediterranean. In this regard, the globalization of the economy and the consequent disintegration of the traditional rural economies have contributed to coastal urban growth.

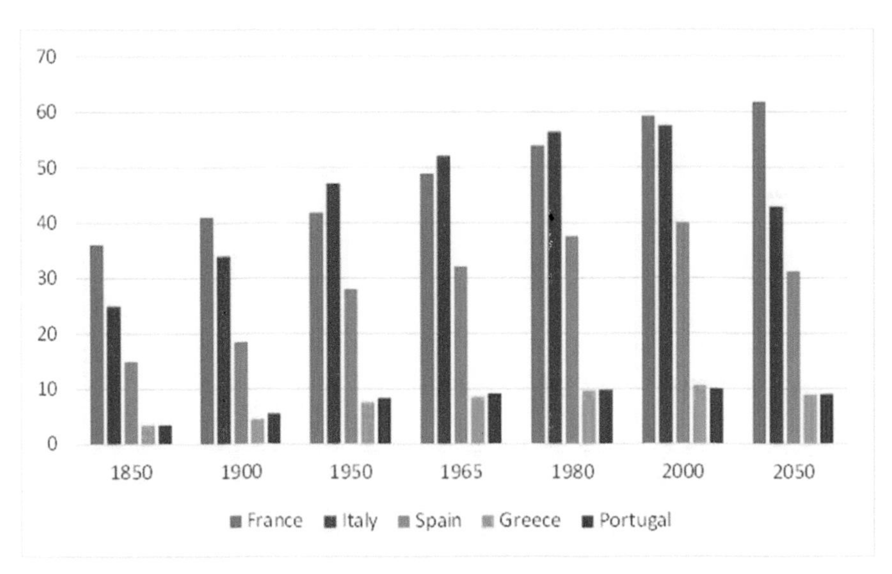

Figure 2.2 Population growth (millions of inhabitants) in the five Mediterranean Countries of EU during the last 150 years.

Source: McEvedy/Jones 1978; UN, 2001.

Table 2.1 Population growth (millions of inhabitants) in the Mediterranean Countries. Mediterranean Europe includes the countries of Figure 2.2. North Africa includes Algeria, Morocco, Tunisia, Lybia and Egypt; Mid-East includes Jordan, Israel, Palestine Authonomy, Lebanon, Syria and Turkey

								Change	
	1850	**1900**	**1950**	**1965**	**1980**	**2000**	**2050**	**1950–2050**	**2000–2050**
Med. Europe	83	103	132.9	150.6	167.3	177.3	154.1	21.2	-23.2
North Africa	13.1	22.3	44.1	62.9	91.4	142.8	239.4	193.3	96.6
Mid-east	10	13	20.8	31.1	44.4	55.7	98.8	78.0	43.1

Source: Authors' elaboration from UN (2006) data.

In this context of population change, urbanization trends around the Mediterranean basin have differed significantly. In Southern Europe the rate of urbanization has been projected to increase by 2030. An even greater trend is expected in North Africa. According to the United Nations Urbanization Prospects, by 2030, projections estimate that the urban population will be 84.5% in Spain, 82.2% in France, 81.6% in Portugal, 76.1% in Italy and 71.6% in Greece. These projections clearly show how the pressure for urbanization in Southern European and Northern African cities will increase in the

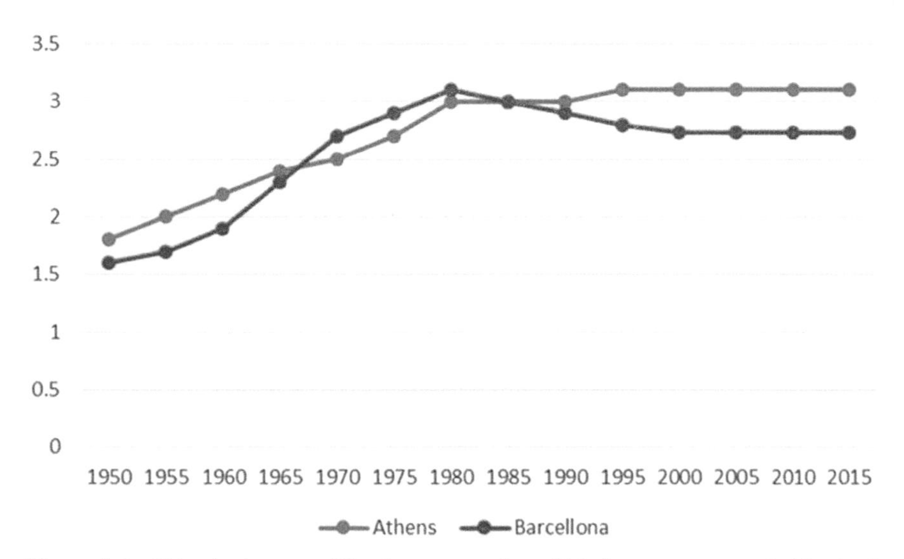

Figure 2.3 Urbanization rate (%) of two examples of Mediterranean coastal cities in the period 1950–2015.

Source: Authors' elaboration from UN (2006) data.

near future. Figure 2.3 shows how, from 1950–2000, two examples of typical Mediterranean coastal cities (i.e., Athens and Barcelona) have experienced a continuous increase in urbanization rates (1.1–1.8-fold) and are expected to stabilize in the next decade up to 2030.

2.3 In between Planned Compaction and Desired Dispersion: The Case of Athens, Greece

The detailed analysis of the urban development of Athens in the 20th century is of interest for various reasons. First, the urban development of Athens can be seen as a typical example of the trajectories shared by many Mediterranean cities, which are changing the urban form paradigm from a compact model to a sprawled and poly-centric one. Therefore, understanding the drivers behind this transition can help to improve our comprehension of Urban Sprawl in the Mediterranean. In this regard, a peculiar characteristic of the Mediterranean the city particularly evident in Athens is the unplanned urban expansion outside city core. Second, Athens is among the few large urban areas of Europe that is still showing a relevant demographic increase. This phenomenon is coupled with a drastic over-urbanization and high rates of land take which are modifying the typical rural landscape. The area under consideration requires a detailed analysis of latent factors and active determinants of the sprawl, which is particularly young and intense, in order to apply appropriate policy strategies to control it. Last, Athens is a very good example for the analysis of the effects of 'Mega-Events' in the new context of Global City-Regions. The Olympic Games of 2004 have made the city an interesting example in Europe for studying how the new 'entrepreneurial city' is influencing urbanization processes.

As in other European contexts such as Spain and Italy, Athens' economic structures in the aftermath of the World War II were not based on manufacturing but rather on urbanization economies which have triggered an industrialisation process. However, this process did not result in the deindustrialisation/disurbanization phase that happened after the 1970s in Northern Europe (Economou et al., 2007). Athens' urban development mainly occurred through illegal housing, sprawling onto suburban land with a lower cost but severe infrastructure deficiencies including lack of sewage systems and poor infrastructures (Nickayin et al., 2020). This led to multiple adverse consequences on the natural capital. Furthermore, it was associated with the increase of social exclusion and vulnerability of the residents of these fringe areas.

The morphological structure of the Greek landscape has influenced the traditional urban system of Athens, which was characterized by a clear polarisation between city and countryside. Indeed, during the 20th century, this relationship has rapidly changed, mainly because of the development of the economic capital, relevant immigration fluxes a lack of effective planning (Tomao et al., 2017; Nickayin et al., 2020). In this context, the progressive urbanization of the surrounding areas of greater Athens has caused the spatial distribution of the activities and therefore the disappearance of the urban-rural polarity.

Athens had to face also with one of the highest population densities in Mediterranean Europe (about 20,000 inhabitants per km^2 in the consolidated urban core). The city structure and the relationships with the surrounding areas have to be adapted. The increasing request for new sites for the construction of commercial, industrial, residential and recreational assets has led to the expansion of the urban area outside its traditional boundaries. The more suitable locations to this aim were Messoghia plain and Thriassian. To better understand the sprawl process in Attica, it is worth considering that the functional urban area of Athens can be described through four prefectures: Athens, Piraeus, East and West Attica. Athens and Pireo are the more densely populated and contain most of the industrial sites of the Attica region, while East and West Attica host most of the rural territories, including the Messoghia and Thiriassio plains.

The first phase of the urban expansion of Athens (1850s–1900s) was characterised by a balanced distribution of the population over the territory and an equal growth of urban and rural areas. However, since the 1860s growth rates of the centre of Athens began to increase faster than in the rest of Attica. In the second phase (1900s–1940s) a compact poly-centric configuration was developed around the two main urban centres of Athens and Piraeus. In this period, new industrial activities were attracted in the surroundings of Athens and especially in Piraeus, where the commercial port is located. As a direct consequence of this, the rates of population growth sharply increased in these two areas. The demand for new housing for the new residents caused the establishment of new spontaneous suburbs around the centres of Athens and Piraeus. These informal settlements, even if not exactly characterized by a low density, can be considered the first signs of Urban Sprawl.

During the third phase (1950s–1980s) of urban expansion, the traditional poly-centrism of the region was transformed. Workers and rural immigrants moved to the fringe areas of Athens and Piraeus in search of affordable housing close to the industries. Consequently, the residential densities in the urban centres became more and more similar to those in the peripheries. Moreover, in these decades the process of the urbanization of the Messoghias plain also began. According to the results of various studies about land consumption in the Attica region, during the period from 1960 to 1980, new urbanized areas were established at the urban-rural interface on agriculture and the surrounding forested lands (Tomao et al., 2017). This urban-rural depolarization was favoured by the permissive urban policies and building code together with the development of transport infrastructures. The sprawl process has been made evident by the demographic trend of population migration from the main urban centres towards the rest of Attica.

In the 1980s–1990s decades, the fourth phase of sprawl has been described and is characterized by the movement of middle-class and elite from the urban centres. Athens' periphery (i.e. Acharnes, Ano Liossia, Filis), rural zones (Messoghia and Thriassio plains) and coastal areas experienced rapid urban growth in those years. Following this period, in the decade 1990–2000, the natural capital in the rural areas was threatened by the abandonment of cultivated fields and the conversion of forests to pastures as a consequence of recurrent fires. In both cases, these land-use changes were promoted by suburbanization and then Urban Sprawl. The conclusion on the demographic and urban development of the region is that the traditional city-countryside polarisation of the Greek territory has radically been replaced by sprawl patterns throughout the 20th century.

In the new millennium, the 2004 Olympic Games have greatly influenced the dynamics of sprawl in Attica. Urban competitiveness for natural and financial resources has grown more than ever, favouring the creation of an entrepreneurial city model. In this context, such Mega-Events have represented an opportunity for self-promotion and therefore have been preceded by infrastructure development and strategies for promoting the image of Athens as the hosting city (i.e. upgrade of the underground and suburban railway, expansion of the International airport and improved connection of the archaeological sites). Therefore, the Olympic Games acted as a sort of 'catalyst' for the new policies oriented to the improvement of the urban reality of Athens. However, the Games have also caused the expansion of the functional urban limits due to the emergence of new spatial links and raising values of rural and peripheral land. Furthermore, urban interventions have been carried out in the entire Athens' urban area, determining a 'multi-nucleated urban regeneration program'. As a consequence, new sprawl patterns and therefore the 'spillover' of Athens have been indirectly favoured (Salvati and Serra, 2016).

In the case of the Attica region, the so-called Post-Olympic leapfrogging Urban Sprawl was the most common sprawl pattern. It is characterized by the expansion of urban areas through sub-centres around the Olympic sites spatially separated by the city core (even if connected by roads infrastructure). Other examples of leapfrog development in Attica are 'Metropolitan Periphery Industrial Development' (urban clusters around industrial sites) and the 'Housing Cooperatives' (new settlements created by groups of interest). Ribbon sprawl patterns have been detected throughout the region. Strip urban developments have sprang along major transport axes and such resultis proof that Infrastructure-driven Urban Sprawl is dominant or widely diffused in Greece (and the Mediterranean).

2.4 Concluding Remarks

Along with the European coastal regions, Urban Sprawl and landscape change are becoming endemic. The environmental impacts of sprawl are evident in a number of ecologically sensitive areas located in coastal zones and mountain areas. With regards to the Area studied by the Plan Bleu and Centre d'Activités Régionales (2020) there is little prospect of relief over the next two decades, especially with a predicted increase in population. The development-related impacts on coastal ecosystems and their habitats and services, have produced major changes in these coastal zones. The Mediterranean coast, one of the world's 34 biodiversity hotspots, is particularly affected, and the increased demand for water for urban use, competes with irrigation water for agricultural land. The consequences of Urban Sprawl on the environment, economy and quality of the cities include:

- the ever-growing cost of urban infrastructures;
- the loss of farming and natural land along the coasts that is often amongst the richest in the countries in question;
- the disappearance of wetlands and coastal erosion;
- the destruction of highly valuable natural habitats (shallow water areas, posidonia beds, sand dunes, turtle nesting sites, etc.);
- reduction of small-scale fishing;
- global degradation of the quality of Mediterranean landscapes;
- intensification of the effects of natural disasters in urbanised areas.

The list above refers specifically to the consequences of Urban Sprawl along the coasts of the Mediterranean. This list can be further expanded if we also include the observed impacts of low-density urbanization in the rest of the region, which include:

- a constant increase in demand for travel, linked to the decoupling of home from work;
- generalised congestion along the main transport arteries and, consequently, a drop in travel speed;
- recurrent shortcomings in public transport provision in terms of servicing and level of service;
- a constant rise in greenhouse gasses emissions linked to the transport sector, mainly road transport which is heavily dependent on fossil energy;
- increased land consumption per capita;
- higher dependency on fossil fuels;
- loss of economic attractiveness of the city.

Faced with these trends, the policies of integrated management of the coastal areas, of sustainable conservation of the coastline, wetlands and periurban farming land, promoting sustainable agriculture and rural development of inland areas and of integrating tourism and sustainable development should be strengthened everywhere. The Mediterranean Commission on Sustainable

Development endeavours to produce ideas and strategic proposals in this regard and efforts are being made on the regional level and in most of the Mediterranean countries, but results are still very limited.

In these regards, processes driving urban growth are inherently related to multiple socioeconomic factors, making the analysis of urban form vs socioeconomic functions a challenging and complicated endeavor (Bruegmann, 2005; Cohen, 2006; Hall and Pain, 2006; Angel et al., 2011). To identify the main determinants of city growth, it is necessary to consider a number of fundamental factors and contextual indicators, including economic and demographic variables, the socio-spatial structure and territorial patterns, as well as institutional and cultural attributes (Musterd and Ostendorf, 1998; Kazepov, 2005; Cassiers and Kesteloot, 2012).

In Mediterranean cities, long-established urban contexts remain heavily influenced by a complex interplay between these factors (Morelli and Salvati, 2010; Salvati et al., 2013; Salvati and Di Feliciantonio, 2014). Burgel (1975) depicted the Mediterranean region as represented – overall – by innumerable landscapes, urban forms, economic structures and social contexts, which would justify the inherent difficulty of explaining and interpreting the uniqueness of each Mediterranean city within this complex panorama. The case study of Athens demonstrates the variety that these areas have, despite the homogenous Mediterranean context. They differ in morphology and have had a different territorial and political configuration, but have common land use trajectories (Tomao et al., 2021a; 2021b). The study of these cities simultaneously is useful for addressing the issues of the management and policies of sprawl.

References

Angel, S., Parent, J., Civco, D. L., Blei, A., and Potere, D. (2011). The dimensions of global urban expansion: Estimates and projections for all countries, 2000–2050. *Progress in Planning*, 75(2), pp. 53–107.

Arapoglou, V. P., and Sayas, J. (2009). New facets of urban segregation in southern Europe: Gender, migration and social class change in Athens. *European Urban and Regional Studies*, 16(4), pp. 345–362.

Arbaci, S. (2007). Ethnic segregation, housing systems and welfare regimes in Europe. *European Journal of Housing Policy*, 7(4), pp. 401–433.

Arbaci, S., and Malheiros, J. (2013). De-segregation, peripheralisation and the social exclusion of immigrants: Southern European cities in the 1990s. In Linking Integration and Residential Segregation (pp. 67–96). *Routledge*.

Bell, S., Peña, A. C., and Prem, M. (2013). Imagine coastal sustainability. *Ocean & Coastal Management*, 83, pp. 39–51.

Bruegmann, R., 2005. Sprawl: a compact history. University of Chicago Press, Chicago.

Burgel, G. (2004). Athènes, de la balkanisation à la mondialisation. Méditerranée, 103(3), pp. 59–63.

Cassiers, T., and Kesteloot, C. (2012). Socio-spatial inequalities and social cohesion in European cities. *Urban Studies*, 49(9), pp. 1909–1924.

Christiansen, P., and Loftsgarden, T. (2011). Drivers behind urban sprawl in Europe. TØI report, 1136, 2011.

Cohen, Barney. (2006). Urbanization in Developing Countries: Current Trends, Future Projections, and Key Challenges for Sustainability. *Technology and Society* 28(1-2), 63–80.

Colantoni, A., Grigoriadis, E., Sateriano, A., Venanzoni, G., and Salvati, L. (2016). Cities as selective land predators? A lesson on urban growth, deregulated planning and sprawl containment. *Science of the Total Environment*, 545, pp. 329–339.

Couch, C., Petschel-held, G., Leontidou, L., 2007. Urban Sprawl In Europe: Landscapes, Land-use Change and Policy. *Blackwell*, London.

De Muro, P., Monni, S., and Tridico, P. (2011). Knowledge-based economy and social exclusion: shadow and light in the roman socio-economic model. *International Journal of Urban and Regional Research*, 35(6), pp. 1212–1238.

Di Feliciantonio, C., and Salvati, L. (2015). 'Southern'Alternatives of Urban Diffusion: Investigating Settlement Characteristics and Socio-*Economic Patterns in Three Mediterranean Regions*. Tijdschrift voor economische en sociale geografie, 106(4), pp. 453–470.

Economou, D., Petrakos, G. and Psycharis, Y. (2007). National urban policy in Greece. In L. Van den Berg, E. Braun and J. Van der Meer (eds.) *National Policy Responses to Urban Challenges in Europe* (193–216). Aldershot, UK: Ashgate.

Gasparella, L., Tomao, A., Agrimi, M., Corona, P., Portoghesi, L., and Barbati, A. (2017). Italian stone pine forests under Rome's siege: learning from the past to protect their future. *Landscape Research*, 42(2), pp. 211–222.

Hall, P., Pain, K., (2006). The polycentric metropolis. Learning from megacity regions in Europe. London, UK: Earthscan.

Kasanko, M., Barredo, J. I., Lavalle, C., McCormick, N., Demicheli, L., Sagris, V., and Brezger, A. (2006). Are European cities becoming dispersed?: A comparative analysis of 15 European urban areas. *Landscape and Urban Planning*, 77(1-2), pp. 111–130.

Kazepov, Y. (2005). Cities of Europe: Changing Contexts, Local Arrangements, and the Challenge to Urban Cohesion. Oxford, UK: Blackwell.

King, R., Proudfoot, L. and Smith, B. (1997). The mediterranean. *Environment and Society*. London: Arnold.

Kurt, S. (2016). Analysis of Temporal Change Taking Place at the Coastline and Coastal Area of the South Coast of the Marmara Sea. *Gaziantep University Journal of Social Sciences*, 15(3).

Leontidou, L. (1990). The Mediterranean city in transition. *Cambridge University Press*, Cambridge.

Leontidou, L. (1996). Alternatives to modernism in (southern) urban theory: Exploring in-between spaces. *International Journal of Urban and Regional Research*, 20(2), pp. 178–195.

Malheiros, J. (2002). Ethni-cities: residential patterns in the Northern European and Mediterranean metropolises–implications for policy design. *International Journal of Population Geography*, 8(2), pp. 107–134.

Maloutas, T. (2007). Socio-economic classification models and contextual difference: The 'European Socio-economic Classes'(ESeC) from a South European angle. *South European Society & Politics*, 12(4), pp. 443–460.

Mascarenhas, A., Haase, D., Ramos, T. B., and Santos, R. (2019). Pathways of demographic and urban development and their effects on land take and ecosystem services: The case of Lisbon Metropolitan Area, Portugal. *Land Use Policy*, 82, pp. 181–194.

Mikhaylov, A. S., Mikhaylova, A. A., and Kuznetsova, T. Y. (2018). Coastalization effect and spatial divergence: Segregation of European regions. *Ocean & Coastal Management*, 161, pp. 57–65.

Morelli, V. G., and Salvati, L. (2010). Ad hoc urban sprawl in the Mediterranean city: Dispersing a compact tradition?. Edizioni Nuova Cultura.

Musterd, S. and Ostendorf, W. (1998). Urban segregation and the welfare state: Inequality and exclusion in western cities. London, UK: *Routledge*.

Nickayin, S. S., Tomao, A., Quaranta, G., Salvati, L., and Gimenez Morera, A. (2020). Going toward Resilience? Town Planning, Peri-Urban Landscapes, and the Expansion of Athens, Greece. *Sustainability*, 12(24), pp. 10471.

Pagliarin, S. (2018). Linking processes and patterns: Spatial planning, governance and urban sprawl in the Barcelona and Milan metropolitan regions. *Urban Studies*, 55(16), pp. 3650–3668.

Pili, S., Grigoriadis, E., Carlucci, M., Clemente, M., and Salvati, L. (2017). Towards sustainable growth? A multi-criteria assessment of (changing) urban forms. *Ecological Indicators*, 76, pp. 71–80.

Salvati, L. and Di Feliciantonio, C. (2014). 'Exploring social mixite in the urban context through a simplified diversity index.' *Current Politics and Economics of Europe* 24(3–4), pp. 1–11.

Salvati, L., Morelli, V. G., Rontos, K., and Sabbi, A. (2013). Latent exurban development: City expansion along the rural-to-urban gradient in growing and declining regions of southern Europe. *Urban Geography*, 34(3), pp. 376–394.

Salvati, L., and Gargiulo Morelli, V. (2014). Unveiling Urban Sprawl in the Mediterranean Region: Towards a Latent Urban Transformation?. *International Journal of Urban and Regional Research*, 38(6), pp. 1935–1953.

Salvati, L. and Forino, G. (2014). A'laboratory'of landscape degradation: social and economic implications for sustainable development in peri–urban areas. *International Journal of Innovation and Sustainable Development*, 8(3), pp. 232–249.

Salvati, L., Quatrini, V., Barbati, A., Tomao, A., Mavrakis, A., Serra, P., Sabbi, A., Merlini, P., and Corona, P. (2016). Soil occupation efficiency and landscape conservation in four Mediterranean urban regions. *Urban Forestry and Urban Greening*, 20, pp. 419–427.

Salvati, L. and Serra, P. (2016). Estimating rapidity of change in complex urban systems: A multidimensional, local-scale approach. *Geographical Analysis*, 48(2), pp. 132–156.

Schneider, A. and Woodcock, C. E. (2008). Compact, dispersed, fragmented, extensive? A comparison of urban growth in twenty-five global cities using remotely sensed data, pattern metrics and census information. *Urban Studies*, 45(3), pp. 659–692.

Sreeja, K. G., Madhusoodhanan, C. G., and Eldho, T. I. (2016). Coastal zones in integrated river basin management in the West Coast of India: Delineation, boundary issues and implications. *Ocean & Coastal Management*, 119, pp. 1–13.

Tomao, A., Quatrini, V., Corona, P., Ferrara, A., Lafortezza, R., and Salvati, L. (2017). Resilient landscapes in Mediterranean urban areas: Understanding factors influencing forest trends. *Environmental Research*, 156, pp. 1–9.

Tomao, A., Mattioli, W., Fanfani, D., Ferrara, C., Quaranta, G., Salvia, R., & Salvati, L. (2021a). Economic Downturns and Land-Use Change: A Spatial Analysis of Urban Transformations in Rome (Italy) Using a

Geographically Weighted Principal Component Analysis. Sustainability, 13(20), 11293.

Tomao, A., Quaranta, G., Salvia, R., Vinci, S., & Salvati, L. (2021b). Revisiting the 'southern mood'? Post-crisis Mediterranean urbanities between economic downturns and land-use change. Land Use Policy, 111, 105740.

UN (United Nations). (2006). World Urbanization Prospects: The 2005 Revision. New York: United Nations.

United Nations Environment Programme/Mediterranean Action Plan and Plan Bleu (2020). *State of the Environment and Development in the Mediterranean*. Nairobi.

United Nations. (2019). World Population Prospects 2019, Online Edition. New York, NY: *Department of Economic and Social Affairs*. Retrieved from https://population.un.org/wpp/

Valev, E. B. (2009). The problems of development and interaction of seaside territories in Europe. *Regional Studies*, 1(22), pp. 11–23.

Zambon, I., Serra, P., Sauri, D., Carlucci, M., and Salvati, L. (2017). Beyond the 'Mediterranean city': Socioeconomic disparities and urban sprawl in three Southern European cities. Geografiska Annaler: Series B, *Human Geography*, 99(3), pp. 319–337.

3

Wildfires and the Local Context: An Empirical Analysis of a Peri-Urban District

Giuseppe Cillis[1], Rosa Coluzzi[2], Antonio Tomao[3], Maria Lanfredi[2], and Luca Salvati[4]

[1]Scuola di Scienze Agrarie, Forestali, Alimentari e Ambientali, University of Basilicata, Viale dell'Ateneo Lucano, I-85100 Potenza, Italy
[2]Institute of Methodologies for Environmental Analysis of the Italian National Research Council (IMAA–CNR), Contrada Santa Loja snc, I-85050 Tito Scalo, Italy
[3]Department for Innovation in Biological, Agro-food and Forest systems (DIBAF), University of Tuscia, Via S. Camillo de Lellis, I-01100, Viterbo, Italy
[4]Department of Methods and Models for Economics, Territory and Finance (MEMOTEF), Faculty of Economics, Sapienza University of Rome, Via del Castro Laurenziano 9, I-00161 Rome, Italy
E-mail: giuseppe.cillis@unibas.it; rosa.coluzzi@imaa.cnr.it; antonio.tomao@unitus.it; maria.lanfredi@imaa.cnr.it; luca.salvati@uniroma1.it

Abstract

Attica region (Greece) is one of the Mediterranean areas mostly affected by significant fire danger due mainly to unregulated urban sprawl.

This chapter investigates the spatial pattern of 881 forest fires reported in 59 municipalities within this area in the years 2009–2012.

By directly using six fire indicators and eight spatial context indicators under a multivariate analysis framework, the idea that a given local-scale fire (whose profile is characterized through severity, density, and selectivity in land use) is connected to a specific set of land and socioeconomic variables was investigated.

The study revealed two crucial domains for both size and selectivity of forest fires and demographic factors associated with the urban-rural gradient and mean income (socioeconomic background). In the first analysis domain, forest and grazed burnt land and fire density appeared not correlated with any socioeconomic variable. Moreover, some fire-related variables appeared also uncorrelated with municipality altitude and average declared income. On the other hand, in the second analysis domain, a positive correlation was identified between the total area of each municipality and the city centre distance with the percentage of burned surface per municipality, the average size of fires and proportion of cultivated land burned. In addition, these last fire variables are negatively correlated with population growth and density and the proportion of dense settlements. The final results confirm that the spatial distribution of forest fires in Attica is influenced by the urban-rural gradient and simultaneously the socioeconomic status of the resident population is an uninfluential variable in this relationship.

Keywords: Socio-demographic changes; Declared income; Indicators; Multivariate analysis; Fire risk; Mediterranean region.

3.1 Introduction

The effect of some biophysical gradients, such as distance from the sea and elevation, influences the landscape structure. In vulnerable Mediterranean regions this condition is amplified by the history of these environments, modelled by the millennial action of mankind and by the potential effects of climate change in terms of irregular patterns of rainfall seasonality and increasing of extreme events (*e.g.*, Greco et al., 2018; Coluzzi et al., 2020; Lanfredi et al., 2020). In these areas, the expansion and spatial distribution of urban settlements have historically shaped the surrounding landscape in a more fruitful way (Polyzos et al., 2008; Kaniewski et al., 2013; Salvati et al., 2016a; Cuadrado-Ciuraneta et al., 2017; Duvernoy et al., 2018; Zambon et al., 2018; Inostroza et al., 2019). In recent years there has been an increase in urban sprawl taking the form of discontinuous, low-density settlements (Van de Voorde et al., 2011; Serra et al., 2014; Salvati et al., 2016b; Salvati and Serra, 2016; Carlucci et al., 2017; Zambon et al., 2017; Lemoine-Rodríguez et al., 2020) even in those southern European cities that have often been regarded as models of compact, dense, land-saving cities (Telesca et al., 2009; Hennig et al., 2015; Marraccini et al., 2015; Guastella et al., 2019).

In addition to this, the development of Mediterranean territories is influenced by the interaction between direct causes and latent drivers of ecological and socio-economic nature (García-Ruiz et al., 2013; Ibáñez et al., 2015; Ferrara et al., 2016; Pinto Correia et al., 2018; Salvati et al., 2018; Salvati et al., 2019; Zambon and Salvati, 2019). A typical example is the phenomenon of land degradation. This is an extremely complex issue which generally determines negative environmental implications resulting in the disruption of ecological balances and the loss of ecosystem services in terms of reduced soil productivity and fertility (Satriani et al., 2012; Imbrenda et al., 2013), and also detrimental economic impacts (Sutton et al., 2016; Yan et al., 2016; Pacheco et al., 2018).

In this analysis, additional elements were considered such as landscape characteristics, geographic region of reference, spatial scale and spatial coverage justifying the key role assumed by social sciences in this kind of study (Rubio and Recatalá, 2006; Lambin and Geist, 2007; Abu Hammad and Tumeizi, 2012). In addition, some landscapes could have limited resilience due to a possible difficulty in reacting to external stresses, although they have been often subject to transformations and alterations linked to natural dynamics (Blum, 2013; Kelly et al., 2015; Kosmas et al., 2016).

In a perspective of sustainable development, human activity is considered responsible for the 'net degradation' of environments (Wilson and Juntti, 2005; Johnson and Lewis, 2007; Briassoulis, 2011; Imbrenda et al., 2014; Zdruli, 2014; D'Emilio et al., 2018; Egidi et al., 2020). However, monitoring actions are difficult to identify due to the high diversification of land degradation processes, thus limiting planning activities at different spatial and administrative levels (Kosmas et al., 2014; Brandt and Geeson, 2015). Although this issue is complicated to address, in recent years there has been a renewed interest for this, especially in the biogeographic region of the Mediterranean Sea (Bajocco et al., 2012; Colantoni et al., 2015; Lanfredi et al., 2015; Delfanti et al., 2016).

In several studies and projects funded by the European Community, land degradation has been addressed by just focusing on its complexity. This concept, which implicitly proposes a relationship with the path of sustainable land development (Salvati and Zitti, 2012), has been described as a process of quantitative and qualitative decline over time of the natural capital (Montanarella, 2007; Kairis et al., 2014; D'Odorico and Ravi, 2016; Ferrara et al., 2020). The relationship between land degradation and sustainable development appears still underexplored in Mediterranean regions even if scientific literature is plenty of valuable contributions (Wilson and Juntti, 2005; Coluzzi et al., 2019; Halbac-Cotoara-Zamfir et al., 2020).

The World Commission on Environment and Development (WCED) in 1987 has produced the so-called Bruntland Report 'Our Common Future', in which the concept of 'Sustainability' is firstly defined. In this report, sustainable development is described as: 'a development that meets the needs of the present without compromising the ability of future generations to satisfy their own'. The concept of sustainability is wide. It includes both issues related to the economic flow of income between generations and the maintenance of the quality and quantity of natural goods whose existence can be altered to a significant degree by anthropic activity. However, the question of 'future generations' is quite vague and therefore generalizes the definition of sustainability. The problem is mainly related to the determination of the time span of political, socio-economic, ethical and strategic decisions that affect the persistence and conservation of landscape and environmental heritage (Scoones, 2016). For this reason, subjects and entities organized at different levels (from local to international) represent the reference point for tackling these problems, as the actions of individual subjects have a short-term time horizon and thus appear less effective.

The relationship between sustainability and desertification has raised a strong debate both from the environmental and the socio-economic point of view (Salvati and Zitti, 2007; Salvati and Zitti, 2008; Salvati and Zitti, 2009). In this work, this link was explored by addressing the essential issues of interpretative nature under a common framework. In a given model of sustainable development, the processes of land degradation can be implemented as a contribution to environmental degradation (Zonn et al., 2017), thus providing informative support for spatial indicators useful for regional and national monitoring. In addition, the interpretative framework, today widely disseminated at the EU level, firstly includes the environmental policies proposed in 1977 by the United Nations in the World Convention for the Fight against Drought and Desertification (UN, 1977).

Regarding the planning of peri-urban landscapes, soil erosion and wildfire are threat factors that are becoming increasingly impactful due to the expansion of suburban sprawl. In fact, rural areas surrounding large cities are more fragmented in terms of land cover and land use classes, making these landscapes more vulnerable to the factors mentioned above. In particular, large forest fires, considering the characteristics of the environment, are the most impactful factor as they can radically transform the structure, configuration, conformation and diversity of landscape patterns (Syphard et al., 2019). In addition, fires can cause several consequences damaging also the social and economic structure of territory by triggering also soil degradation processes

(Coluzzi et al., 2007; Salvati and Bajocco, 2011; Karamesouti et al., 2015; Kosmas et al., 2016; Alonso-Sarría et al., 2016; Symeonakis et al., 2016; Lanorte et al., 2019).

The starting hypothesis, on which this work is based, considers that a fire on a local scale is associated with a certain set of socio-economic and geographic variables as well as typical fire attributes (density, severity and land use). The study area analyzed is the Attica peninsula (Greece) which is characterized by fire risk and significant anthropic pressure. Within the municipalities of this Mediterranean region, 881 forest fires were recorded during the time span 2009–2012 on which a geospatial distribution analysis was performed. By adopting a multivariate approach that takes into account six fire and eight contextual indicators, we investigated the spatial distribution of the identified forest fires to provide decision-makers with suitable and timely strategies involving peri-urban and forest areas.

3.2 Experimental Design

3.2.1 Study Area

The area chosen as a case study (Figure 3.1) covers about 3000 km^2 and is located in mainland Greece (Attica region) including the capital Athens. From a morphological point of view, the region is quite varied and contains several mountains that exceed 1000 m above sea level (Mount Parnitha 1413 m, Pateras 1132 m, Penteli 1109 m and Hymettus 1026 m). Outside the most populated and urbanized areas, there are three coastal plains (Thriasio, Messoghia and Marathon, see Salvati et al., 2012). Scarce territorial planning controls have indirectly caused unrestricted urban sprawl without socio-economic and environmental impacts assessment (Pili et al., 2017; Gounaridis et al., 2019; Tomao et al., 2017; Nickayin et al., 2020). From an administrative point of view, the region is divided into several municipalities including the island of Salamina. This subdivision is the result of the 'Kallikatris law' that in 2010 has reformed the local governement in Greece.

3.2.2 Data and Variables

To create the forest fire database, records compiled by different fire departments were used assuming their reliability and completeness. For each event, the following information was considered: description of the burned place (included any useful information for geolocation as road and toponym), the municipalities whose territory is affected by the ignition, date and probable

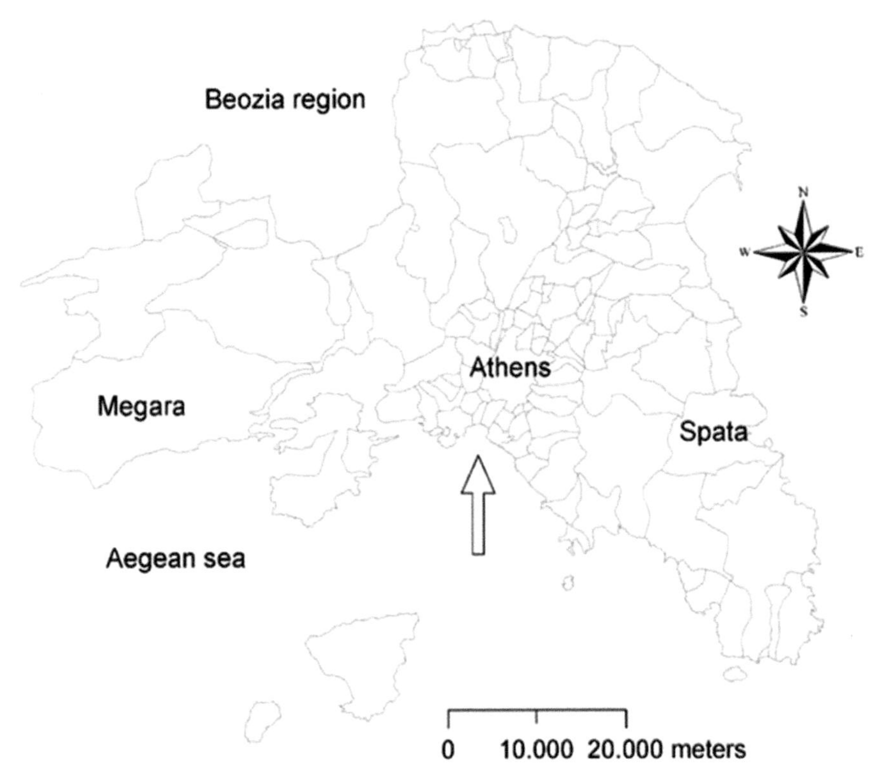

Figure 3.1 A map illustrating the administrative structure of the study area.
Source: Authors' elaboration on official maps.

hour of ignition and area of fire estimated through in situ inspections. The whole database includes 881 forest fire events with related descriptive attributes that occurred in the 59 municipalities of the study area recorded between 2009 and 2012. All data enabled each fire to be associated with one of the 59 municipalities in the region.

The database also includes the causes of ignition, most of which have an anthropic origin and only a minimum part is due to natural calamitous events as sometimes happens in Mediterranean regions. However, the fire department records (for this area and study period) are not so specific to identify for each fire the true causes. Considering the dates of fire events, the period with the highest number of registrations falls between late spring and early autumn, *i.e.*, in typically warmer and drier periods. It was also possible to include in the database the land cover class affected by the fire.

Table 3.1 Descriptive statistics of fires occurred in the study area during the 4-year period

Variable	2009	2010	2011	2012	Total
Number of fires	184	304	194	199	881
Burnt area (km^2)	199.3	17.9	3.8	28.2	249.1
Average fire size (ha)	108.3	5.9	1.9	14.2	28.3
Woodland (%)	50.2	25.0	68.6	55.8	49.3
Cropland (%)	16.8	29.4	3.4	21.1	18.0
Pastures (%)	25.4	41.7	24.2	22.9	26.3

In order to make the interpretation of data easier, three land cover classes were considered: 1. forest and natural habitats including shrublands, garrigue and evergreen scrub, 2. cultivated land and 3. pastures. Finally, with regard to the density of wildfires, the value recorded in the period 2009–2012 is 0.29 events/km^2 and each event has an average size of about 28 ha (Table 3.1).

Generally, as revealed by a case-specific study proposed by Bajocco and Ricotta (2008), in this area the statistics of the size of occurred fires is well defined; in fact, only a small part (1%) has a size greater than 100 ha, as most fires (80%) do not exceed the extent of 1 ha. To characterize fire season, it is also necessary to take into account the climatic conditions recorded during the examined period. The data provided by the National Observatory of Athens (NOAA – Thissio Station) define the year 2009 as relatively humid and the year 2010 as extraordinarily dry and hot while the others (2011–2012) show standard climatic conditions. Therefore, the study period is characterized by a mixed climate pattern. Regarding regime and severity of fire events, in the studied period there were: in 2009 few but remarkably severe fires; in 2010 moderate severity of fire events (both for forest area and total area involved); in 2011 low fire severity and reduced number of events; in 2012 intermediate fire profiles.

3.2.3 Wildfire Indicators

The methodology used is based on the implementation of indicators accounting for both fire-related factors (density, severity and selectivity) and properties of the urban-rural gradient (demographics, spatial patterns, mean income and population density). In total, six fire indicators and eight socio-economic parameters were calculated at the municipal level (Table 3.2). The data used for processing were retrieved from the Hellenic Statistical Authority, the Ministry of Finance and the Ministry of Environment. The spatial distribution of the selected indicators is shown in Figure 3.2.

Table 3.2 List of variables considered in this work

Acronym	Indicator	Measurement unit	Primary data source
	Wildfires' Variables		
FIRDEN	Density of forest fires on total municipal area (2009–2012)	Fires per km^2	Hellenic Fire Brigade
AVGFSU	Average fire size by municipality (2009–2012)	Hectare	Hellenic Fire Brigade
FIRSU%	Percentage of municipal area affected by fires (2009–2012)	Burnt area/km^2	Hellenic Fire Brigade
WOOD	Proportion of burnt forests on total burnt area (2009–2012)	%	Hellenic Fire Brigade
CROP	Proportion of burnt cropland on total burnt area (2009–2012)	%	Hellenic Fire Brigade
PAST	Proportion of burnt pastures on total burnt area (2009–2012)	%	Hellenic Fire Brigade
	Background Variables		
DENPOP	Population density (2011)	Inhabitants/km^2	Hellenic Statistic Authority
DIST	Distance from Athens	km	Hellenic Statistic Authority
VARDEM	Annual population growth rate (2001–2011)	%	Hellenic Statistic Authority
POPRAT	The ratio of present to resident population (2011)	%	Hellenic Statistic Authority
ADJBUI	The ratio of adjacent buildings on total buildings (2001)	%	Hellenic Statistic Authority
SUPMUN	Municipality surface area	km^2	Ministry of the Environment
ELEV	Elevation (0: lowland; 1: upland municipalities)	Dummy	Ministry of the Environment
INCOME	Average per-capita declared income (2011)	Euros	Ministry of Finance

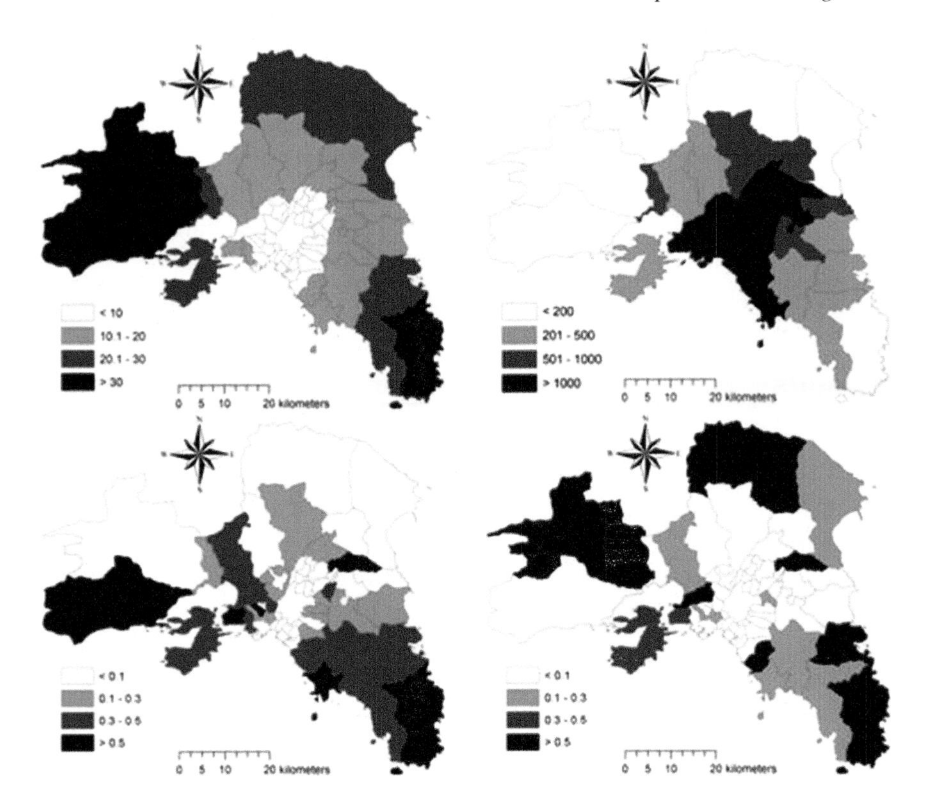

Figure 3.2 Maps showing some features of the analyzed area (upper left: average distance of each municipality from Athens' center [km]; upper right: population density [inhabitants/km^2]; lower left: density of forest fires [per year and km^2], lower right: percentage of burnt land on total municipal area [%]).

Source: Authors' elaboration on official data.

3.2.4 Data Analysis

The spatial unit of analysis adopted in this study, by integrating information from different statistical and field surveys, enables the assessment of changes in the spatial distribution of population along the urban gradient. Official municipal limits (source Hellenic Statistical Office, 2012), excluding those in which no fire was recorded, were used to build the maps necessary to investigate the spatial distribution of each indicator. To identify the main components that shape the fire profile and the socioeconomic and land characteristics of local municipalities in the Attica region, two distinct Principal Component Analyses were performed. Two distinct graphs of factor scores and factor loadings were produced from the extraction of factors with eigenvalues greater than 1. In order to compare the distribution of six fire with eight socioeconomic indicators in the investigated municipalities, a non-parametric Spearman rank correlation analysis by testing at $p < 0.05$ after Bonferroni correction for multiple comparisons was performed.

As the last analysis, to detect similarities in the spatial distribution of the 14 indicators, a non-hierarchical cluster analysis was also developed by a non-Euclidean metric based on Ward's agglomeration rule.

3.3 Results

From the analysis of fires recorded within the municipalities studied, it emerges that in those municipalities (20) with high population pressure, with few urban green spaces (excluding open gardens and abandoned lands), no fires were recorded. Therefore, events were observed only in 39 out of the total number of municipalities. The distribution of wildfire density recorded (Figure 3.2) is spatially varied. Small fires (the most numerous) are concentrated mainly in peri-urban municipalities with high accessibility. On the other hand, if we consider as a parameter the percentage of burnt area with respect to the total municipal surface, a different pattern can be observed: fewer rural (and less accessible) municipalities host severe and often more frequent fires. Therefore, the results show a relationship between fires and urban context. In fact, the data show that there are different profiles of fires emerging at a local scale according to the socio-economic conditions found along the urban gradient.

The relevance of different interpretative dimensions at a local level was evident when analysing contextual indicators outlining the spatial opposition between typical forest fires and fires events affecting agricultural areas and pastures with municipalities organized primarily along this gradient. Small-size fires occurred more frequently in woodlands with respect to large-size fires that more commonly affected cropland and pasture areas.

In the study area, two prevalent dimensions of spatial arrangement emerged: population growth and density effectively describe the urban-rural gradient, correctly indicated by population density and increasing distance from the centre city of Athens. Socioeconomic distinctions at the urban level reflect the divergence between poor and rich municipalities even on the basis of their altitude above sea level. Municipalities located in mountain areas and close to the main urban centres are more occupied by middle and high social classes.

Table 3.3 Non-parametric Spearman rank correlation coefficients between forest fires parameters and socioeconomic variables (only significant correlations at $p < 0.05$ after Bonferroni's correction for multiple comparisons are reported)

Variable	DENPOP	DIST	VARDEM	POPRAT	ADJBUI	SUPMUN
AVGFSU	−0.55	0.48	−0.45	–	−0.44	0.49
FIRSU%	−0.55	0.51	–	0.47	–	0.45
CROP	−0.78	0.78	−0.79	0.52	−0.78	0.73

Table 3.3 shows the results of the correlation analysis between fire and socioeconomic variables performed using Spearman's rank coefficient matrix. The model emerging suggests a very complex pattern based on the severity of fire events and the corresponding selectivity of the land use involved. The analysis shows that socioeconomic variables are not correlated with pasture and forest burned areas and fire density. In addition, fire variables were not correlated with some variables associated with municipalities (elevation class and average income). On the contrary, correlation is evident among other factors. Indeed, the statistical survey shows a negative correlation between the ratio of burnt area per municipality, the ratio of agricultural land burned and the average size of fires with population size and density, the ratio of living population to the resident population and the proportion of dense settlements. Lastly, the correlation is positive with the overall area of each municipality and the distance from the city centre.

One of the results inferred from the analysis is the possibility to distinguish the fires recorded in the rural area surrounding the urban perimeter and the fires that occurred in the neighborhood of Athens. The first category of fires is characterized by low selectivity, high severity and high density and involves mainly certain types of burnable areas (prevalence of agricultural land and pastures).

3.3.1 Discussing Changes in 'Wildfires' Regime in the Study Area

The empirical findings of our study indicate a close relationship between the spatial distribution of forest fires and urban-rural gradient in the Attica region. As for the socioeconomic component, the influence of the resident population on this relationship is negligible. Landscape and territory represent two very complex issues to be dealt with and, above all, to be managed as their dynamics are linked to different variables, encompassing both the natural and the human ones (Figure 3.3). Therefore, their analysis requires

Figure 3.3 Abandoned land with tree covering in Athens' fringe.
Source: Authors' photographic archive.

a multidisciplinary approach combining environmental aspects with socio-economic dynamics to explain the evolution and predict future scenarios (Antrop and Van Eetvelde, 2017). In addition to this, the issue of the management of non-renewable land resources should be considered. This way, to avoid competition in land use and the occurrence of land degradation phenomena, it is necessary to follow appropriate land use planning and land resource management approaches (Gasparella et al., 2017; Ramamurthy, 2018).

Contemporary approaches are always focused on one or more specific purposes and are closely linked to the principles of efficiency, equity and sustainability. To achieve these goals, participatory and integrated approaches must be applied to manage land resources at different levels (van Dam et al., 2010). In this context, it is critical to apply methodologies and techniques for landscape analysis that are as complete and broad as possible (Maxwell et al., 2019; Koç and Yılmaz, 2020). Alongside the techniques and features to be incorporated into landscape analyses, it is crucial to deepen assessments of these features by considering how spatial and temporal differences influence landscape transformation dynamics (Imbrenda et al., 2018; Picuno et al., 2019).

Thus, the interdisciplinary nature of landscape studies, the different definitions associated with it imply that contrasting methods for landscape assessment can be found in the literature and there is no single method that succeeds in characterizing all spatial and temporal levels without including trade-offs within the analysis (Simensen et al., 2018). On the other hand, land degradation also requires a multidisciplinary approach for detailed knowledge of the fundamental features that characterize the land (Quaranta et al., 2020; Xie et al., 2020). In fact, land degradation can be studied using several methods, such as direct field observation and remote sensing. Compared to the field method, the integration of geographical information systems, geospatial analysis and remote sensing is much more economical and time-efficient as it is possible to study a large area (hundreds of square kilometers) by leveraging small amounts of digital data (Bonfiglio et al., 2002; El Baroudy, 2011; Pignatti et al., 2015). In addition, this approach provides models for the assessment and study of scenarios that can be replicated in different territories and improved if needed (Dubovyk, 2017).

The analysis of impacts on the environment and its dynamics necessarily benefits from the use of tools for mapping and management of geographic data. In addition, the ability to extract information from 'big data' technologies, has allowed the implementation of techniques and methodologies with a high level of completeness and that facilitate analysis at both global and local scale (Lokers et al., 2016). Moreover, the improvement of these tools also guarantees greater interoperability between different data management systems and therefore allows for homogenization operations of different qualitative and quantitative data even spatially and temporally different (Cillis et al., 2019). These methods enable the development of geo-statistic indicators useful to determine trajectories of change and the corresponding drivers (Zhang et al., 2018). The scientific literature reports many cases of combined use of simple and complex indicators and spatial systems to support decisions in the field of environmental planning; but the integration of ecological and socio-economic issues within these schemes still needs further investigations (Salvati and Zitti, 2012).

The integrity of landscapes being transformed by rapid ecological dynamics needs real-time data on environmental fluxes to investigate the influence of determinants versus natural endogenous evolution in the perspective to control ecosystem pathways (Plieninger et al., 2015). A thorough investigation of the interactions between biophysical patterns and socioeconomic factors is essential to delineate land degradation and the evolution of its main drivers (Manh Vu et al., 2014). In view of this, to experimentally examine environmental processes whose knowledge may be biased by the observation scale, it is necessary 1. to address issues at different temporal and spatial scales and 2. to prefer comparative studies over single case studies. This is because environmental dynamics and land transformation are based on phenomena operating at distinct scales (local and/or global) and with different intensities (Estes et al., 2018).

Local, regional, and country levels are the adopted geographic scales. These subdivisions are consistent with the degree of spatial detail of the chosen parameters and are easily interpretable for stakeholders since these scales ensure the study of changes in environmental indicators and ecosystem services across different geographical dimensions (Hein et al., 2006). Regional differences in the availability of natural resources depend on how processes act at the geographical level of interest. In fact, environmental phenomena, influenced by the spatial scale, imply a different impact on land resources since the involved socio-economic and biophysical drivers act with different intensities on the analysed landscapes (Liu et al., 2017). Moreover, the issue of equitable distribution of resources is highly scale-dependent in environmental dynamics. Thus, assessing these confluence processes is critical to promoting sustainable resource management. In fact, many social conflicts over the use of land assets can originate from areas neighbouring the hot-spots of environmental degradation, even though these zones are not directly affected by land degradation (Salvati et al., 2011). Therefore, implementing policy reforms, with a spatial scale sufficiently congruent to reduce this type of conflict, ensures more effective sustainable and geographically balanced development (Sala et al., 2015).

Suitable methodologies in evaluating the impact of desertification phenomena have recently been developed by analyzing public policies at all levels of intervention. The ongoing crisis of conventional governance systems and economic insecurity is making this issue more and more prominent. In fact, in developed economies tackling land degradation has been demonstrated to be crucial because companies have greater awareness and interest in the effectiveness of public policy and so they act with greater pressure on the decision-making process (Halbac-Cotoara-Zamfir et al., 2020). In contrast, citizens living in the drylands of Mediterranean regions seem to be ruled by insecurity and uncertainty, compromising the effectiveness and applicability of environmental policies (Barbero-Sierra et al., 2015). In the review of potential effects of policies, it is necessary to consider key questions influencing the actions of land stakeholders who have a role in land degradation triggering/exacerbating (Reynolds et al., 2011).

The first issue is that a specific impact on the land is not directly and necessarily linked to specific policy actions. The second refers to the fact that the impacts of a policy may be evident or not. The third issue is at odds with the classical and logical approach of policy strategies because it concerns the dynamic nature of the desertification phenomenon and its high variability (Hogl et al., 2016; Sietz et al., 2017). Direct impacts on the environment can also result from strategies that do not contemplate policy-making, non-decisional procedures and no action formulation. This dynamic has emerged from many case studies although not sufficiently recognized by literature, so the problem of the so-called costs of inaction also needs to be addressed scientifically (Alberini et al., 2016; McConnell and Hart, 2019). Hence, as a specific action is implemented, the full practice of policy strategy should not be considered as a net measure but as a fuzzy set of generalized decision making on several areas simultaneously (Werdiningtyas et al., 2020).

The integration of several strategies to address different objectives represents the most useful approach to dealing with such issues. This has been highlighted in the scientific literature which advises against addressing the problem in a sectoral manner and differentiated by disciplines. The latter approach has been used by political governments in recent years; however, although the commitment has been important and substantial, it needs to be changed. The actions to deal with and combat land degradation should be implemented on the basis of contextual assessments and considering multiple factors; as, for example, it has been addressed in recent years - even in national planning – by adopting the DPSIR scheme. This methodology is based on an integrated perspective (Webb et al., 2017) to obtain demonstrable and measurable outcomes on target groups of territorial actors and stakeholders.

With appropriate experimental assessment, the effect of occurred changes on the environment can be identified. In fact, it can be defined through the visible effects resulting from the surveys, or through the impact of the policy strategy as the overall impact on the environment (Escadafal et al., 2015). Based on the abovementioned DPSIR, public interventions can be implemented in terms of multiple solutions, needing coordination at suitable geographical scales. Specifically, measures acting directly on the agricultural sector and, to a lower degree, on the industrial one, are among the policies that can become potential tools for encouraging sustainable development and contributing positively to socio-economic cooperation. This can be paradigmatically observed in Southern Europe, where geographical differences persist between the north and south of the countries. Development policies can act on two different levels (Salvati et al., 2014). The first concerns the development of strategies that should be oriented towards a sort of 'total production' through the horizontal integration of production (agro-silvo-pastoral management); whereas the second is focused on the vertical integration of the earth products following the different stages of transformation and marketing, in the perspective to optimize investments in an economic context where land assures low productivity.

3.4 Concluding Remarks

Many case studies located in Mediterranean regions have shown that among factors triggering fire risk the spatial pattern of land use and socio-demographic components are the most important. Specifically, population concentration, the load of grazing, intensification of agricultural activities, land abandonment and urban sprawl appear as the most influencing factors. This theory also partially emerges from this work where a close relationship was identified between socio-economic parameters and peculiar fire profiles (land use selectivity and severity) in a Mediterranean peri-urban environment. It has been demonstrated that it is possible to infer information on the fires' patterns and their features from the analysis of socio-economic and demographic indicators by adopting appropriate exploratory statistical tools. This statistical analysis was carried out on the basis of variables processed from official and freely available databases according to the logic of open data. The structure of the variables implemented represents a useful tool that can be integrated into a decision support system for fire risk monitoring. Moreover, they can be spatialized, mapped and interconnected using other types of geographical data such as administrative boundaries, biogeographical classes, physiographic layers, etc. In addition, this approach ensures the possibility of integration within the information system of data derived from on-site surveys or remote sensing platforms. Finally, planning at the local scale and for small administrative areas (municipalities or districts) can take advantage of the findings of this investigation. For example, they can be used to implement fire plans that include monitoring, control and intervention measures in the most vulnerable areas and to suggest sustainable forest management practices in the wildland-urban interface.

References

Abu Hammad, A. and Tumeizi, A. (2012). Land degradation: socioeconomic and environmental causes and consequences in the eastern Mediterranean. *Land Degradation & Development*, 23(3), 216–226.

Alberini, A., Bigano, A., Post, J., and Lanzie, E. (2016). Approaches and issues in valuing the costs of inaction of air pollution on human health. OECD environment working papers, No. 108, Paris: OECD Publishing.

Alonso-Sarría, F., Martínez-Hernández, C., Romero-Díaz, A., Cánovas-García, F., and Gomariz-Castillo, F. (2016). Main environmental features leading to recent land abandonment in Murcia region (Southeast Spain). *Land Degradation & Development*, 27(3), 654–670.

Antrop, M. and Van Eetvelde, V. (2017). Landscape perspectives: The holistic nature of landscape. Springer: Dordrecht, The Netherlands, 2017; ISBN 978-94-024-1183-6.

Bajocco, S., and Ricotta, C. (2008). Evidence of selective burning in Sardinia (Italy): which land-cover classes do wildfires prefer? *Landscape Ecology*, 23(2), 241–248.

Bajocco, S., De Angelis, A., and Salvati, L. (2012). A satellite-based green index as a proxy for vegetation cover quality in a Mediterranean region. *Ecological Indicators*, 23, 578–587.

Barbero-Sierra, C., Marques, M. J., Ruiz-Pérez, M., Escadafal, R., and Exbrayat, W. (2015). How is desertification research addressed in Spain? Land versus soil approaches. *Land Degradation & Development*, 26(5), 423–432.

Blum, W. E. (2013). Soil and land resources for agricultural production: general trends and future scenarios-a worldwide perspective. *International Soil and Water Conservation Research*, 1(3), 1–14.

Bonfiglio, A., Cuomo, V., Lanfredi, M., and Macchiato, M. (2002). Interfacing NOAA/ANHRR NDVI and soil truth maps for monitoring vegetation phenology at a local scale in a heterogeneous landscape of Southern Italy. *International Journal of Remote Sensing*, 23(20), 4181–4195.

Brandt, J. and Geeson, N., (2015). Desertification indicator system for Mediterranean Europe. In monitoring and modelling dynamic environments (eds A.P. Dykes, M. Mulligan and J. Wainwright).

Briassoulis, H. (2011). Governing desertification in Mediterranean Europe: The challenge of environmental policy integration in multi-level governance contexts. *Land Degradation & Development*, 22(3), 313–325.

Carlucci, M., Grigoriadis, E., Rontos, K., and Salvati, L. (2017). Revisiting a hegemonic concept: Long-term 'Mediterranean urbanization'in between city re-polarization and metropolitan decline. *Applied Spatial Analysis and Policy*, 10(3), 347–362.

Cillis, G., Statuto, D., and Picuno, P. (2019). Historical maps processed into a GIS for the assessment of forest landscape dynamics. Proceedings of the Public Recreation and Landscape Protection—With Sense Hand in Hand, 180–184.

Colantoni, A., Ferrara, C., Perini, L. and Salvati, L. (2015). Assessing trends in climate aridity and vulnerability to soil degradation in Italy. *Ecological Indicators* 48, 599–604, doi:10.1016/j.ecolind.2014.09.031.

Coluzzi, R., di Donna, I., Lanorte, A., and Lasaponara, R. (2007, October). Mapping forest fuel types by using satellite ASTER data and neural nets. In Remote sensing for agriculture, ecosystems, and hydrology IX (6742, p. 67420T). *International Society for Optics and Photonics.*

Coluzzi, R., D'Emilio, M., Imbrenda, V., Giorgio, G. A., Lanfredi, M., Macchiato, M., Ragosta, M., Simoniello, T., and Telesca, V. (2019). Investigating climate variability and long-term vegetation activity across heterogeneous Basilicata agroecosystems. *Geomatics, Natural Hazards and Risk*, 10(1), 168–180.

Coluzzi, R., Fascetti, S., Imbrenda, V., Italiano, S. S. P., Ripullone, F., and Lanfredi, M. (2020). Exploring the Use of Sentinel-2 Data to Monitor Heterogeneous Effects of Contextual Drought and Heatwaves on Mediterranean Forests. Land, 9(9), 325.

Cuadrado-Ciuraneta, S., Durà-Guimerà, A., and Salvati, L. (2017). Not only tourism: Unravelling suburbanization, second-home expansion and "rural" sprawl in Catalonia, Spain. *Urban Geography*, 38(1), 66–89.

D'Odorico, P., and Ravi, S. (2016). Land degradation and environmental change. *Biological and Environmental Hazards, Risks, and Disasters.* Netherlands: Elsevier, 219–227.

D'Emilio, M., Coluzzi, R., Macchiato, M., Imbrenda, V., Ragosta, M., Sabia, S., and Simoniello, T. (2018). Satellite data and soil magnetic susceptibility measurements for heavy metals monitoring: findings from Agri Valley (Southern Italy). *Environmental Earth Sciences*, 77(3), 1–7.

Delfanti, L., Colantoni, A., Recanatesi, F., Bencardino, M., Sateriano, A., Zambon, I., and Salvati, L. (2016). Solar plants, environmental degradation and local socioeconomic contexts: A case study in a Mediterranean country. *Environmental Impact Assessment Review* 2016, 61, 88–93, doi:10.1016/j.eiar.2016.07.003.

Dubovyk, O. (2017). The role of Remote Sensing in land degradation assessments: opportunities and challenges. *European Journal of Remote Sensing*, 50(1), 601–613.

Duvernoy, I., Zambon, I., Sateriano, A., and Salvati, L. (2018). Pictures from the other side of the fringe: Urban growth and peri-urban agriculture in a post-industrial city (Toulouse, France). *Journal of Rural Studies*, 57, 25–35.

Egidi, G., Salvati, L., Cudlin, P., Salvia, R., and Romagnoli, M. (2020). A new 'Lexicon'of land degradation: Toward a holistic thinking for complex socioeconomic issues. *Sustainability*, 12(10), 4285.

El Baroudy AA (2011) Monitoring land degradation using remote sensing and GIS techniques in an area of the middle Nile Delta, Egypt. *Catena* 87(2), 201–208

Escadafal, R., Marques, M. J., Stringer, L. C., and Akhtar-Schuster, M. (2015) Opening the door to policy relevant, Interdisciplinary research on land degradation and development. *Land Degradation & Development* 26, 409–412.

Estes, L., Elsen, P. R., Treuer, T., Ahmed, L., Caylor, K., Chang, J., Choi, J. J., and Ellis, E. C. (2018). The spatial and temporal domains of modern ecology. *Nature Ecology & Evolution* 2, 819–826.

Ferrara, A., Kelly, C., Wilson, G.A., Nolè, A., Mancino, G., Bajocco, S., Salvati, L. (2016). Shaping the role of "fast" and "slow" drivers of change in forest-shrubland socio-ecological systems. *Journal of Environmental Management* 169, 155–166, doi:10.1016/j.jenvman.2015.12.027.

Ferrara, A., Kosmas, C., Salvati, L., Padula, A., Mancino, G., and Nolè, A. (2020). Updating the MEDALUS-ESA framework for worldwide land degradation and desertification assessment. *Land Degradation and Development*, https://doi.org/10.1002/ldr.3559.

García-Ruiz, J. M., Nadal-Romero, E., Lana-Renault, N., and Beguería, S. (2013). Erosion in mediterranean landscapes: changes and future challenges. *Geomorphology*, 198, 20–36.

Gasparella, L., Tomao, A., Agrimi, M., Corona, P., Portoghesi, L., and Barbati, A. (2017). Italian stone pine forests under Rome's siege: learning from the past to protect their future. *Landscape Research*, 42(2), 211–222.

Gounaridis, D., Chorianopoulos, I., Symeonakis, E., and Koukoulas, S. (2019). A Random Forest-Cellular Automata modelling approach to explore future land use/cover change in Attica (Greece), under different socio-economic realities and scales. *Science of the Total Environment*, 646, 320–335.

Greco, S., Infusino, M., De Donato, C., Coluzzi, R., Imbrenda, V., Lanfredi, M., Simoniello, T., and Scalercio, S. (2018). Late spring frost in Mediterranean beech forests: extended crown dieback and short-term effects on moth communities. *Forests*, 9(7), 388.

Guastella, G., Oueslati, W., and Pareglio, S. (2019). Patterns of urban spatial expansion in European cities. *Sustainability*, 11(8), 2247.

Halbac-Cotoara-Zamfir, R., Smiraglia, D., Quaranta, G., Salvia, R., Salvati, L., and Giménez-Morera, A. (2020). Land Degradation and Mitigation Policies in the Mediterranean Region: A Brief Commentary. *Sustainability*, 12(20), 8313.

Hein, L., Van Koppen, K., De Groot, R. S., and Van Ierland, E. C. (2006). Spatial scales, stakeholders and the valuation of ecosystem services. *Ecological Economics*, 57(2), 209–228.

Hennig, E. I., Schwick, C., Soukup, T., Orlitová, E., Kienast, F., and Jaeger, J. A. (2015). Multi-scale analysis of urban sprawl in Europe: Towards a European de-sprawling strategy. *Land Use Policy*, 49, 483–498.

Hogl, K., Kleinschmit, D., and Rayner, J. (2016). Achieving policy integration across fragmented policy domains: Forests, agriculture, climate and energy. Environment and Planning C: *Government and Policy*, 34(3), 399–414.

Ibáñez, J., Contador, J. L., Schnabel, S., Fernández, M. P., and Valderrama, J. M. (2014). A model-based integrated assessment of land degradation by water erosion in a valuable Spanish rangeland. *Environmental Modelling & Software*, 55, 01–213.

Imbrenda, V., Coluzzi, R., Lanfredi, M., Loperte, A., Satriani, A., and Simoniello, T. (2018). Analysis of landscape evolution in a vulnerable coastal area under natural and human pressure. *Geomatics, Natural Hazards and Risk*, 9(1), 1249–1279.

Imbrenda, V., D'Emilio, M., Lanfredi, M., Ragosta, M., and Simoniello, T. (2013). Indicators of land degradation vulnerability due to anthropic factors: tools for an efficient planning. In Sustainable Practices: Concepts, Methodologies, Tools, and Applications (pp. 1400–1413). IGI Global.

Imbrenda, V., D'Emilio, M., Lanfredi, M., Macchiato, M., Ragosta, M., and Simoniello, T. (2014). Indicators for the estimation of vulnerability to land degradation derived from soil compaction and vegetation cover. *European Journal of Soil Science*, 65(6), 907–923.

Inostroza, L., Hamstead, Z., Spyra, M., and Qureshi, S. (2019). Beyond urban–rural dichotomies: Measuring urbanisation degrees in central

European landscapes using the technomass as an explicit indicator. *Ecological Indicators*, 96, 466–476.

Johnson, D. L., and Lewis, L. A. (2007). Land degradation: creation and destruction. Rowman and Littlefield.

Kairis, O., Kosmas, C., Karavitis, C., Ritsema, C., Salvati, L., Acikalin, S., ... and Ziogas, A. (2014). Evaluation and selection of indicators for land degradation and desertification monitoring: types of degradation, causes, and implications for management. *Environmental Management*, 54(5), 971–982.

Kaniewski, D., Van Campo, E., Morhange, C., Guiot, J., Zviely, D., Shaked, I., ... and Artzy, M. (2013). Early urban impact on Mediterranean coastal environments. *Scientific Reports*, 3(1), 1–5.

Karamesouti, M., Detsis, V., Kounalaki, A., Vasiliou, P., Salvati, L., Kosmas, C. (2015). Land-use and land degradation processes affecting soil resources: Evidence from a traditional Mediterranean cropland (Greece). Catena 132, 45–55, doi:10.1016/j.catena.2015.04.010.

Kelly, C., Ferrara, A., Wilson, G. A., Ripullone, F., Nolè, A., Harmer, N., Salvati, L. (2015). Community resilience and land degradation in forest and shrubland socio-ecological systems: Evidence from Gorgoglione, Basilicata, Italy. *Land Use Policy*, 46, 11–20.

Koç, A., and Yılmaz, S. (2020). Landscape character analysis and assessment at the lower basin-scale. *Applied Geography*, 125, 102359.

Kosmas, C., Kairis, O., Karavitis, C., Ritsema, C., Salvati, L., Acikalin, S., ... and Ziogas, A. (2014). Evaluation and selection of indicators for land degradation and desertification monitoring: methodological approach. *Environmental Management*, 54(5), 951–970.

Kosmas, C., Karamesouti, M., Kounalaki, K., Detsis, V., Vassiliou, P., Salvati, L. (2016). Land degradation and long-term changes in agro-pastoral systems: An empirical analysis of ecological resilience in Asteroussia - Crete (Greece). Catena 147, 196–204, doi:10.1016/j.catena.2016.07.018.

Lambin, E. F. and Geist, H. J. (2007). Causes of land use and land cover change. Washington DC: Encyclopedia of Earth, *Environmental Information Coalition, National Council for Science and the Environment.*

Lanfredi, M., Coluzzi, R., Imbrenda, V., Macchiato, M., and Simoniello, T. (2020). Analyzing space–time coherence in precipitation seasonality across different European climates. *Remote Sensing*, 12(1), 171.

Lanfredi, M., Coppola, R., Simoniello, T., Coluzzi, R., D'Emilio, M., Imbrenda, V., and Macchiato, M. (2015). Early identification of land degradation hotspots in complex bio-geographic regions. *Remote Sensing*, 7(6), 8154–8179

Lanorte, A., Cillis, G., Calamita, G., Nolè, G., Pilogallo, A., Tucci, B., and De Santis, F. (2019). Integrated approach of RUSLE, GIS and ESA Sentinel-2 satellite data for post-fire soil erosion assessment in Basilicata region (Southern Italy). *Geomatics, Natural Hazards and Risk*.

Lemoine-Rodríguez, R., Inostroza, L., and Zepp, H. (2020). The global homogenization of urban form. An assessment of 194 cities across time. *Landscape and Urban Planning*, 204, 103949.

Liu, Y., Bi, J., Lv, J., Ma, Z., and Wang, C. (2017). Spatial multi-scale relationships of ecosystem services: A case study using a geostatistical methodology. *Scientific Reports*, 7(1), 1–12.

Lokers, R., Knapen, R., Janssen, S., van Randen, Y., and Jansen, J. (2016). Analysis of Big Data technologies for use in agro-environmental science. *Environmental Modelling & Software*, 84, 494–504.

Vu, Q. M., Le, Q. B., Frossard, E., & Vlek, P. L. (2014). Socio-economic and biophysical determinants of land degradation in Vietnam: An integrated causal analysis at the national level. *Land Use Policy*, 36, 605–617.

Marraccini, E., Debolini, M., Moulery, M., Abrantes, P., Bouchier, A., Chéry, J. P., ... and Napoleone, C. (2015). Common features and different trajectories of land cover changes in six Western Mediterranean urban regions. *Applied Geography*, 62, 347–356.

Maxwell, A. E., Strager, M. P., Warner, T. A., Ramezan, C. A., Morgan, A. N., and Pauley, C. E. (2019). Large-Area, High Spatial Resolution Land Cover Mapping Using Random Forests, GEOBIA, and NAIP Orthophotography: Findings and Recommendations. *Remote Sensing*, 11(12), 1409.

McConnell, A., and Hart, P. T. (2019). Inaction and public policy: understanding why policymakers 'do nothing'. *Policy Sciences*, 52(4), 645–661.

Montanarella, L. (2007). Trends in land degradation in Europe. In: Sivakumar M.V.K., Ndiang'ui N. (eds) Climate and Land Degradation. Environmental Science and Engineering (Environmental Science). Springer, Berlin, Heidelberg.

Nickayin, S. S., Tomao, A., Quaranta, G., Salvati, L., and Gimenez Morera, A. (2020). Going toward Resilience? Town planning, peri-urban landscapes, and the expansion of athens, Greece. *Sustainability*, 12(24), 10471.

Pacheco, F. A. L., Fernandes, L. F. S., Junior, R. F. V., Valera, C. A., and Pissarra, T. C. T. (2018). Land degradation: Multiple environmental consequences and routes to neutrality. *Current Opinion in Environmental Science & Health*, 5, 79–86.

Picuno, P., Cillis, G., and Statuto, D. (2019). Investigating the time evolution of a rural landscape: How historical maps may provide environmental information when processed using a GIS. *Ecological Engineering*, 139, 105580.

Pignatti, S., Acito, N., Amato, U., Casa, R., Castaldi, F., Coluzzi, R., De Bonis, R., Diani, M., Imbrenda, V., Laneve, G.f, Matteoli, S., Palombo, A., Pascucci, S., Santini, F., Simoniello, T., Ananasso, C., Corsini, G., and Cuomo, V. (2015). Environmental products overview of the Italian hyperspectral prisma mission: The SAP4PRISMA project. In 2015 IEEE International Geoscience and Remote Sensing Symposium (IGARSS), 3997-4000. IEEE.

Pili, S., Grigoriadis, E., Carlucci, M., Clemente, M., and Salvati, L. (2017). Towards sustainable growth? A multi-criteria assessment of (changing) urban forms. *Ecological Indicators*, 76, 71–80.

Pinto Correia, T., Primdahl, J., and Pedroli, B. (2018). European landscapes in transition. Implications for Policy and Practice. Cambridge University Press: Cambridge, UK.

Plieninger, T., Kizos, T., Bieling, C., Le Dû-Blayo, L., Budniok, M. A., Bürgi, M., ... and Verburg, P. H. (2015). Exploring ecosystem-change and society through a landscape lens: recent progress in European landscape research. *Ecology and Society*, 20(2).

Polyzos, S., Christopoulou, O., Minetos, D., and Filho, W. L. (2008). An overview of urban-rural land use interactions in Greece. International *Journal of Agricultural Resources, Governance and Ecology*, 7(3), 276–296.

Quaranta, G., Salvia, R., Salvati, L., Paola, V. D., Coluzzi, R., Imbrenda, V., and Simoniello, T. (2020). Long-term impacts of grazing management on land degradation in a rural community of Southern Italy: Depopulation matters. *Land Degradation & Development*, 31(16), 2379–2394.

Ramamurthy, V. (2018). Trends in land resource management and land use planning. In: Reddy G., Singh S. (eds) Geospatial Technologies in Land Resources Mapping, Monitoring and Management. Geotechnologies and the Environment, vol 21. Springer, Cham.

Reynolds, J. F., Grainger, A., Starord Smith, D. M., Bastin, G., Garcia-Barrios, L., Fernández, R. J., and Verstraete, M. M. (2011). Scientific concepts for an integrated analysis of desertification. *Land Degradation & Development*, 22, 166–183.

Rubio, J. L. and Recatalá, L. (2006). The relevance and consequences of Mediterranean desertification including security aspects. In Kepner W. G., Rubio J. L., Mouat D. A., and Pedrazzini F. (eds) Desertification in the Mediterranean Region. A Security Issue. NATO Security Through Science Series, vol 3. Springer, Dordrecht.

Sala, S., Ciuffo, B., and Nijkamp, P. (2015). A systemic framework for sustainability assessment. *Ecological Economics*, 119, 314–325.

Salvati, L., and Bajocco, S. (2011). Land sensitivity to desertification across Italy: past, present, and future. *Applied Geography*, 31(1), 223–231.

Salvati, L., Ciommi, M. T., Serra, P., and Chelli, F. M. (2019). Exploring the spatial structure of housing prices under economic expansion and stagnation: The role of socio-demographic factors in metropolitan Rome, Italy. *Land Use Policy*, 81, 143–152.

Salvati, L., Quatrini, V., Barbati, A., Tomao, A., Mavrakis, A., Serra, P., Sabbi, A., Merlini, P., and Corona, P. (2016b). Soil occupation efficiency and landscape conservation in four Mediterranean urban regions. Urban For. *Urban Green*, 20, 419–427.

Salvati, L., Mancini, A., Bajocco, S., Gemmiti, R., and Carlucci, M. (2011). Socioeconomic development and vulnerability to land degradation in Italy. *Regional Environmental Change*, 11(4), 767–777.

Salvati, L., and Serra, P. (2016). Estimating rapidity of change in complex urban systems: A multidimensional, local-scale approach. *Geographical Analysis*, 48(2), 132–156.

Salvati, L., Zambon, I., Chelli, F. M., and Serra, P. (2018). Do spatial patterns of urbanization and land consumption reflect different socioeconomic contexts in Europe? *Science of the Total Environment*, 625, 722–730.

Salvati, L. and Zitti, M. (2007). Territorial disparities, natural resource distribution, and land degradation: a case study in southern Europe. *Geojournal*, 70(2), 185–194.

Salvati, L. and Zitti, M. (2008). Regional convergence of environmental variables: Empirical evidences from land degradation. *Ecological Economics*, 68, 162–168.

Salvati, L. and Zitti, M. (2009). Assessing the impact of ecological and economic factors on land degradation vulnerability through multiway analysis. *Ecological Indicators* 9, 357–363, doi:10.1016/j.ecolind.2008.04.001.

Salvati, L. and Zitti, M. (2012). Monitoring vegetation and land use quality along the rural–urban gradient in a Mediterranean region. *Applied Geography*, 32(2), pp. 896–903.

Salvati, L., Zitti, M., and Carlucci, M. (2014). Territorial systems, regional disparities and sustainability: Economic structure and soil degradation in Italy. *Sustainability*, 6(5), 3086–3104.

Salvati, L., Zitti, M., and Perini, L. (2016a). Fifty years on: long-term patterns of land sensitivity to desertification in Italy. *Land Degradation & Development* 27(2), 97–107.

Satriani, A., Loperte, A., Imbrenda, V., and Lapenna, V. (2012). Geoelectrical surveys for characterization of the coastal saltwater intrusion in Metapontum forest reserve (Southern Italy). *International Journal of Geophysics*, 2012.

Scoones, I. (2016). The politics of sustainability and development. *Annual Review of Environment and Resources*, 41, 293–319.

Serra, P., Vera, A., Tulla, A. F., and Salvati, L. (2014). Beyond urban–rural dichotomy: Exploring socioeconomic and land-use processes of change in Spain (1991–2011). *Applied Geography*, 55, 71–81.

Sietz, D., Fleskens, L., and Stringer, L. C. (2017). Learning from non-linear ecosystem dynamics is vital for achieving land degradation neutrality. *Land Degradation & Development*, 28(7), 2308–2314.

Simensen, T., Halvorsen, R., and Erikstad, L. (2018). Methods for landscape characterisation and mapping: A systematic review. *Land Use Policy*, 75, 557–569.

Sutton, P. C., Anderson, S. J., Costanza, R., and Kubiszewski, I. (2016). The ecological economics of land degradation: Impacts on ecosystem service values. *Ecological Economics*, 129, 182–192.

Symeonakis, E., Karathanasis, N., Koukoulas, S., and Panagopoulos, G. (2016). Monitoring sensitivity to land degradation and desertification with the environmentally sensitive area index: The case of lesvos island. *Land Degradation & Development*, 27(6), 1562–1573.

Syphard, A. D., Rustigian-Romsos, H., Mann, M., Conlisk, E., Moritz, M. A., and Ackerly, D. (2019). The relative influence of climate and housing development on current and projected future fire patterns and structure loss across three California landscapes. *Global Environmental Change*, 56, 41–55.

Telesca, L., Coluzzi, R., and Lasaponara, R. (2009). Urban pattern morphology time variation in Southern Italy by using Landsat imagery. In Murgante B., Borruso G., and Lapucci A. (eds) Geocomputation and Urban Planning. Studies in Computational Intelligence, Springer, Berlin, Heidelberg, 176, 209–222.

UN, Draft Plan of Action to Combat Desertification, Document A/CONF74/LS2, UN Environment Program, Nairobi, 1977.

Tomao, A., Quatrini, V., Corona, P., Ferrara, A., Lafortezza, R., and Salvati, L. (2017). Resilient landscapes in Mediterranean urban areas: Understanding factors influencing forest trends. *Environmental research*, 156, 1–9.

van Dam, J., Junginger, M., and Faaij, A. P. (2010). From the global efforts on certification of bioenergy towards an integrated approach based on sustainable land use planning. *Renewable and Sustainable Energy Reviews*, 14(9), 2445–2472.

Van de Voorde, T., Jacquet, W., and Canters, F. (2011). Mapping form and function in urban areas: An approach based on urban metrics and continuous impervious surface data. *Landscape and Urban Planning*, 102(3), 143–155.

Webb, N. P., Marshall, N. A., Stringer, L. C., Reed, M. S., Chappell, A., and Herrick, J. E. (2017). Land degradation and climate change: building climate resilience in agriculture. *Frontiers in Ecology and the Environment*, 15(8), 450–459.

Werdiningtyas, R., Wei, Y., and Western, A. W. (2020). Understanding policy instruments as rules of interaction in social-ecological system frameworks. *Geography and Sustainability*, 1(4), 295–303.

Wilson, G. and Juntti, M., (2005) Unravelling desertification: Policies and actor networks in Southern Europe. Wageningen, Holland: Wageningen Academic Publishers 978-90-76998-42-8.

Xie, H., Zhang, Y., Wu, Z., and Lv, T. (2020). A bibliometric analysis on land degradation: Current status, development, and future directions. Land, 9(1), 28.

Yan, F., Zhang, S., Liu, X., Chen, D., Chen, J., Bu, K., ... and Chang, L. (2016). The effects of spatiotemporal changes in land degradation on ecosystem services values in Sanjiang Plain, China. *Remote Sensing*, 8(11), 917.

Zambon, I., Benedetti, A., Ferrara, C., and Salvati, L. (2018). Soil matters? A multivariate analysis of socioeconomic constraints to urban expansion in Mediterranean Europe. *Ecological Economics* 146, 173–183.

Zambon, I. and Salvati, L. (2019). Metropolitan growth, urban cycles and housing in a Mediterranean country, 1910s–2010s. Cities, 95, 102412.

Zambon, I., Serra, P., Sauri, D., Carlucci, M., and Salvati, L. (2017). Beyond the Mediterranean city': socioeconomic disparities and urban sprawl in three Southern European cities. *Geografiska Annaler Series B – Human Geography*, 99(3), 319–337.

Zdruli, P. (2014). Land resources of the Mediterranean: status, pressures, trends and impacts on future regional development. *Land Degradation & Development*, 25(4), 373–384.

Zhang, T. Q., Zheng, Z. M., Lal, R., Lin, Z. Q., Sharpley, A. N., Shober, A. L., ... and Van Cappellen, P. (2018). Environmental Indicator Principium with Case References to Agricultural Soil, Water, and Air Quality and Model-Derived Indicators. *Journal of Environmental Quality*, 47(2), 191–202.

Zonn, I. S., Kust, G. S., and Andreeva, O. V. (2017). Desertification paradigm: 40 years of development and global efforts. *Arid Ecosystems*, 7(3), 131–141.

4

From Deforestation to Forestation: The Long-term Experience of a Mediterranean Area

Antonio Tomao[1], Agostino Ferrara[2], Vito Imbrenda[3], and Luca Salvati[4]

[1]Department for Innovation in Biological, Agro-food and Forest systems (DIBAF), University of Tuscia, Via S. Camillo de Lellis, I-01100, Viterbo, Italy
[2]Scuola di Scienze Agrarie, Forestali, Alimentari e Ambientali, University of Basilicata, Viale dell'Ateneo Lucano, I-85100 Potenza, Italy
[3]Institute of Methodologies for Environmental Analysis of the Italian National Research Council (IMAA–CNR), Contrada Santa Loja snc, I-85050 Tito Scalo, Italy
[4]Department of Methods and Models for Economics, Territory and Finance (MEMOTEF), Faculty of Economics, Sapienza University of Rome, Via del Castro Laurenziano 9, I-00161 Rome, Italy
E-mail: antonio.tomao@unitus.it; agostino.ferrara@unibas.it; vito.imbrenda@imaa.cnr.it; luca.salvati@uniroma1.it

Abstract

The present work analyzes trends over time and spatial patterns of an indicator of efficiency in sustainable governance of woodland (per capita forest land) over a sufficiently long time interval (1960–2010) in a Southern European region (Attica, Greece). The results of our study revealed how forest land per capita in Attica had a general and progressive decrease during the study period. Only a few municipalities showed an increase in the indicator in the most recent decade (2000–2010) as a consequence of forest expansion in abandoned and marginal areas. Results confirm that economically dynamic districts can be sensitive to deforestation, because of land take and urbanization of fringe land. Specific measures such as the protection of forests and rural areas are increasingly needed. The establishment of new forests in rural municipalities and the active management of spontaneous woodlots can help improving their functionality, quality and accessibility, promoting more sustainable urbanization in metropolitan regions.

Keywords: Sustainability, Urban forests, Landscape planning, Socioeconomic context.

4.1 Introduction

The term 'sustainability' is defined by the Bruntland Report drawn up in 1987 by the World Commission on Environment and Development (WCED), as 'a development that meets the needs of the present without compromising the ability of future generations to satisfy their own'. The concept of sustainability, therefore linked to the economic flow of income between generations, expands to include the maintenance of the quality and quantity of the natural heritage. In this regard, urban and peri-urban forests (Konijnendijk et al., 2005) are considered among the most important landscape elements which can improve the quality of the urban environment and therefore the sustainability in the urban context. Indeed, urban forests provide a wide range of ecological benefits including regulation of infiltration and storm water runoff, reduction of greenhouse gases and mitigation of microclimate and air pollution (Dobbs et al., 2014; Escobedo et al., 2019). They also contribute in improving people well-being and restoring cognitive resources (Carrus et al., 2015, Tomao et al., 2016; Tomao et al., 2018; Markwell and Gladwin, 2020; Reyes-Riveros et al., 2021). Even if the importance of natural elements in the urban context is widely recognized, a reduction in their ecological functionality has been observed in several metropolitan regions (Zipperer et al., 2012; Chas-Amil et al., 2013; Bajocco et al., 2016; Xiao et al., 2019) particularly due to forest loss. Indeed, despite forests are expanding in marginal territories as a consequence of spontaneous colonization of abandoned pastures or arable land (Barbati et al., 2013; Cimini et al., 2013; Simoniello et al., 2015; Martín-Forés et al., 2020; Quaranta et al., 2020), disturbances such as wildfires and clearcutting can still contribute in reducing forest cover in peri-urban areas (Chas-Amil et al., 2013). This phenomenon can result in a number of adverse implications linked to land degradation (Bajocco et al., 2011; Salvati et al., 2011; Xie et al., 2020).

Since the 20th century, due to the rapid development of urbanization and industrialization, the deteriorating ecological environment and the increasing use of land resources, land degradation has been increasing and worsening (Hammad et al., 2012; Delfanti et al., 2016; Zambon et al., 2017). In particular, land degradation is threating plant and animal biodiversity (Breg Valjavec et al., 2018), food and energy security (Reed et al., 2011), sustainable socio-economic system development and human living environments (Winslow et al., 2011). Land degradation is also considered among the major causes of land desertification (Kosmas et al., 2014; Ferrara et al., 2020). Moreover, this phenomenon is associated to the reduction of the biological productivity and complexity of several ecosystems such as dry and semi-arid land, grassland, rangeland, forest and wetland, due to a complex interaction of anthropogenic disturbances (e.g., land use change) and ecological drivers (e.g., wind and water erosion) (Smiraglia et al., 2016; Imbrenda et al., 2018; Coluzzi et al., 2019; Xie et al., 2020).

Land degradation across the Mediterranean basin is a clear example of the interaction between ecological and socio-economic immediate causes and latent factors (Bajocco et al., 2012; Imbrenda et al., 2013; Recanatesi et al., 2016; Smiraglia et al., 2019). The great heterogeneity of processes driving land degradation is one of the main issues which complicates monitoring in the perspective of sustainable development and limits the possibility of implementing effective action plans through various response channels (Kairis et al., 2013; Kairis et al., 2015; Karamesouti et al., 2015; Kosmas et al., 2016).

Urban fringe, where human settlements occur in natural or semi-natural landscapes (primarily forests) creating a wildland–urban interface (WUI), is one of the areas most sensitive to land degradation due to the interactions between human and natural processes occurring in the settled area and in the surrounding natural landscape (Radeloff et al., 2005). In this context, although landscapes exhibit some forms of natural modification and deterioration, they may not be able to respond to exogenous stresses, showing limited resilience (Blum, 2013). In this perspective, the degradation of land is seen as mainly caused by anthropogenic factors (Curebal et al., 2015; Salvati and Serra, 2016; Adenle et al., 2020) and therefore of interest in the path of sustainable development (Wilson and Juntti, 2005; Salvati et al., 2012; Kelly et al., 2015).

Understanding factors involved in land degradation processes has stimulated vast scientific literature. At present, most of the research about this topic focuses on the evaluation and monitoring, based on different data sources, of driving factor and on the development of trend simulation and prediction through quantitative models (Xie et al., 2020). Several studies have

Figure 4.1 A typical village embedded into a forest matrix at the fringe: Thrakomakedones, Northern Athens.
Source: Authors' photographic archive.

considered land degradation as a complex and composite concept, which describes how one or more components of natural capital has worsened over time, from a quantitative or qualitative point of view. However, despite relevant contributions (Basso et al., 2000; Wilson and Juntti, 2005; Salvati et al., 2016a), the relationship between sustainable development and land degradation still appears far from a comprehensive understanding, especially in the Mediterranean basin.

In particular, the analysis of the factors that can affect positively and negatively landscape changes in terms of land-use efficiency can be useful for a sustainable management of these areas (Pili et al., 2017; Duvernoy et al., 2018; Masini et al., 2019). In this regard, some indicators describing impact of urbanization on land resources have been developed (see for example Hasse and Lathrop, 2003; Di Feliciantonio and Salvati, 2015; Carlucci et al., 2018; Ciommi et al., 2017; Ciommi et al., 2018; Ciommi et al., 2019; Zambon et al., 2018), but relatively few studies analyzed the interactions between ecological processes and socioeconomic changes on forest peri-urban landscapes (Salvati and Zitti, 2005; Catalán et al., 2008; Salvati and Zitti, 2009; Barbati et al., 2013; Salvati et al., 2013; Tomao et al., 2017).

The present chapter addresses the main problems connected with the definition of an interpretative framework on the theme of environmental sustainability linked to modifications of a landscape surrounding a densely urbanized area. In particular, the guiding idea is to propose an analysis of availability of natural land uses in terms of per-capita forest cover over a sufficiently long time interval (1960–2010) in a Southern European region (Attica, Greece). Athens can be considered an exemplificative case of expanding Mediterranean urban areas, where suburbanization-driven settlement scattering and polycentric development have altered the typical mono-centric spatial organization of several urban regions (Figure 4.1). Furthermore, in the last decades, forest cover of the Attica region has been reduced by human-induced disturbances such as wildfires (755 km^2 were burned between 1983 and 2005, http://oikoskopio.gr/pyroskopio/en/).

4.2 Understanding the Basic Characteristics of the Study Area

The study area coincides with the metropolitan region of Athens (Figure 4.2). The landscape is composed of uplands and mountains. Lowlands are less frequent and are located mainly in the central area of the Athens region. The prevailing climate regime in the area is Mediterranean dry, with an annual temperature of 18°C and mean annual rainfall of 400–500 mm. The region covers approximately 3000 km^2 and is administered by 114 municipal authorities: 56 rural municipalities and 58 urban municipalities. The area is very densely populated and in the last 70 years, population density increased from 1500 inhabitants/km^2 up to 4400 inhabitants/km^2.

After World War II, the area showed a relevant urban expansion. First, the centers of Athens and Piraeus developed as principal urban poles, due to their socioeconomic functions (services and industry, respectively). Then, since the early 1990s the region experienced a process of urban de-concentration and an expansion of discontinuous fringe settlements (Chorianopoulos et al., 2010; Salvati, 2014; Nickayin et al., 2020). More recently, the development of the new urban cores of Mesogaia (centered in the municipality of Markopoulo) and Maroussi, where the main structures of the 2004 Olympic games were established, led to the formation of growing poles outside the

Figure 4.2 Maps of the Athens' metropolitan region showing administrative municipal boundaries in yellow.
Source: Authors' elaboration on Google Earth imagery.

consolidated city (Chorianopoulos et al., 2010; Chorianopoulos et al., 2014; Couch et al., 2007; Di Feliciantonio et al., 2018; Nickayin et al., 2020).

4.2.1 Data Sources and Indicators

As a first step of the analysis, an index of per-capita forest cover (*For*) has been calculated as the ratio between forest area and the resident population of each municipality of the study area. The index has been calculated for each decade in the period from 1960– 2010. Forest area was obtained from the National Statistical Service of Greece (NSSG, 2011, now ELSTAT; www.statistics.gr/en/home/) land-use census for the years 1960, 1970, 1980, 1990 and 2000 and from the Urban Atlas (UA) map referring to 2010 (EEA, 2010). Data of resident population was extracted from the NSSG National Census of Population of the years 1961, 1971, 1981, 1991, 2001 and 2011. Data accuracy was evaluated following Salvati et al. (2012). The reliability of forest land cover data measured for each of the reference years was assessed using independent estimates from additional data as shown in Table 4.1.

Table 4.1 Ancillary data used for data reliability assessment

Ancillary Dataset	Year	Source
Soil map delimitating the urban areas	1948	Inst. Geology Soil Chemistry
LaCoast (LC) project mapping land-use in coastal regions of Europe	1975	Perdigao and Christensen, 2000
Corine Land Cover (CLC) maps	1990 and 2000	EEA, 2006
Pan European GlobCorine map	2009	Salvati, 2014
Municipal data on cropland cover	1961, 1970, 1980, 1990, 1999, 2009	Greek National Census of Agriculture

4.2.2 Statistical Analysis

As shown in Figure 4.3, we first calculated descriptive statistics for the *For* indicator at each considered year. In particular, we have calculated the arithmetic and geometric mean, median, a ratio of median to mean and coefficient of variation. In order to better describe *For* trend over time, municipalities were classified according to recent dynamics of such indicator: 1. positive (For+) or negative (For-) variation in per-capita forest area between 1960 and 2010; 2. time trend in per-capita forest area observed every decade between 1960 and 2010.

4.3 Understanding the Complex Evolution of Forest Cover in the Study Area

Descriptive statistics of *For* in Attica show a clear and progressive decrease in the availability of forest area per inhabitant during the studied period (Figure 4.3). In particular, per-capita forest land (Mean and Median) sharply decreased from 1960 to 2000, stabilising or slightly increasing in the 2000–2010 decade. On the other hand, the coefficient of variation, the asymmetry and the kurtosis increased in the period, indicating a deviation from normality in the statistical distribution and a progressive polarization between municipalities characterized by low and high values of per-capita forest cover.

The spatial distribution of *For* shows how, in 1960, values higher than 1500 m^2/inhabitant could be observed in the municipalities located in the northern part of the region (Figure 4.4). At that time, these territories formed a sort of 'green belt' around the consolidated built-up area of Athens.

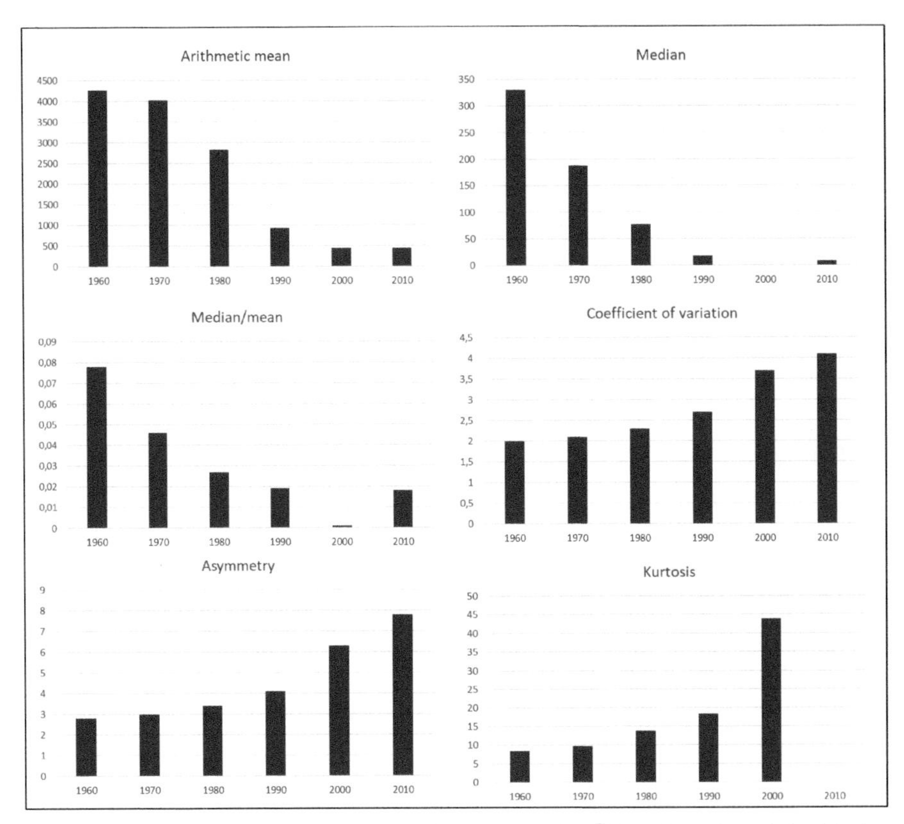

Figure 4.3 Descriptive statistics on per-capita forest land (m^2/inhabitant) in mainland Attica by year. Statistics are calculated using municipalities as the elementary spatial domain.

However, in the more recent decades, this belt has been progressively lost due to the decrease of forest cover. In 2010, the municipalities with a wider forest cover were located in the western and northern part of the region while on the eastern side (Messoghia) most of the rural areas were transformed by a complex dynamic of urbanization and sprawl.

This analysis shows how in the time interval 1960–2010 the complex urbanization process of Athens has affected also forest surfaces. Indeed, only a few municipalities experienced in 2010 *For* values higher than 1500 m^2 per capita. In any case, forest cover does not exceed 50% of the territory of these municipalities.

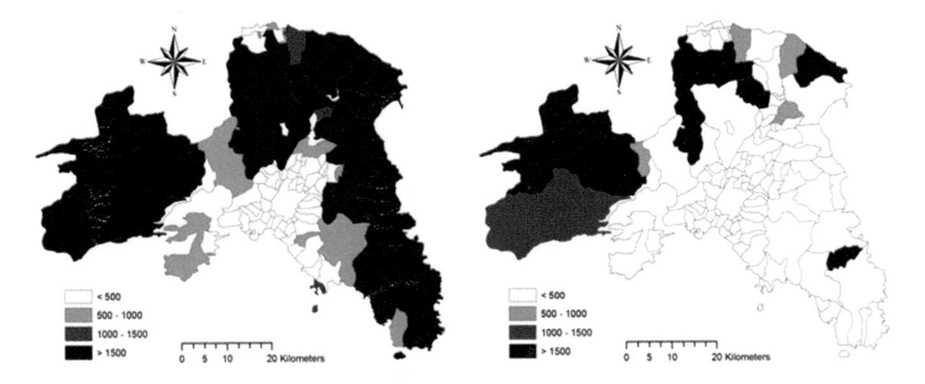

Figure 4.4 Per-capita forest area (m^2) by decade and municipality in Athens' metropolitan area.

Source: Authors' elaboration, redrawn from Tomao et al. (2017).

4.4 Using Per-Capita Forest Dynamics to Profile Municipalities in Attica

Municipalities in Attica have been grouped into six different categories showing homogeneous trends of forest dynamics (Figures 4.5). The percentage of each group is reported in Table 4.2, while the spatial distribution of different trends has been illustrated in Figure 4.6. A total of 28 municipalities showed no changes of *For* in the study period. They are located in the densely

Table 4.2 Municipalities showing similar trends of per-capita forest land in mainland Attica. Mean change of *For* (m^2 per capita) and of forest cover (%) in the 1960–2010 period are also reported.

Group	Number of Municipalities	Land (%)	Change of *For*, 1960–2010	Change of Forest Cover, 1960–2010
no change	28	4.4	0	0
linear negative trend	46	46.9	−1349	−23.6
linear positive trend	7	2.9	22	1.2
inverse U-shaped square trend	7	6.6	−1700	−18.9
U-shaped square trend	19	24.5	−1 700	1,3
non-linear, non-square trend	8	14.8	−6719	−17.1

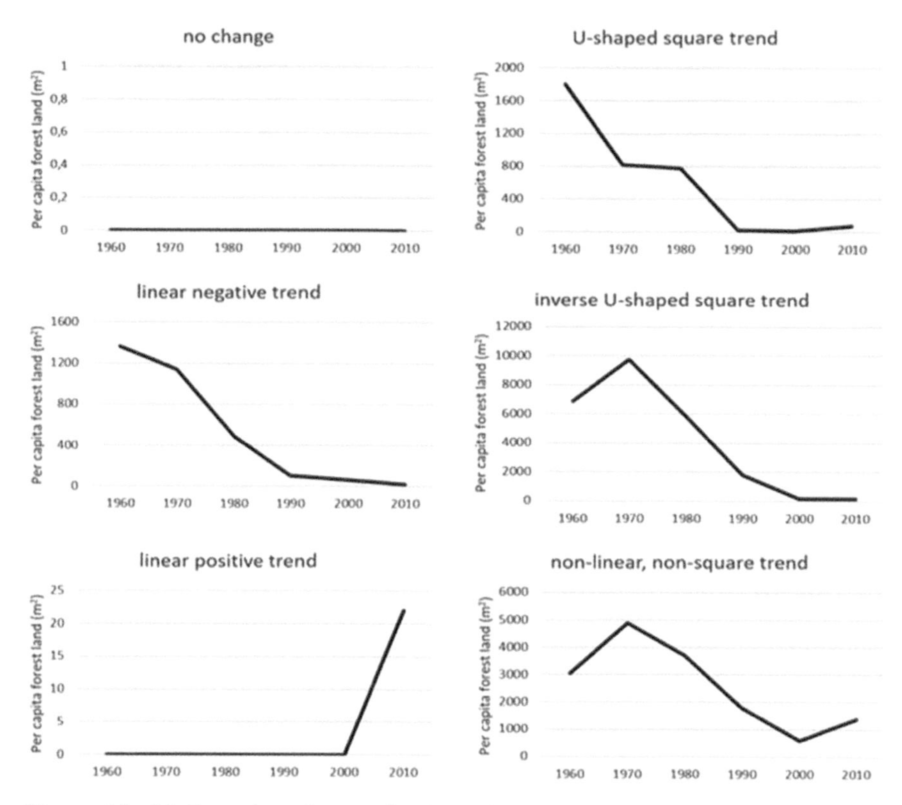

Figure 4.5 Median value of per-capita forest land (*For*) in the municipalities of Attica according to six distinct classification criteria.

urbanized areas of Athens and Pireo, where there are no forest areas, thus resulting in a constant value of 0 for the indicator during period 1960–2010.

The major part of the study area (46.9%, corresponding to 46 municipalities) showed a continuous decreasing trend of *For* which resulted in a reduced availability of forest per inhabitant of 1349 m^2 on average. Only 7 municipalities (2.9% of the study area) showed a positive trend. They are located in the northern part of the region and within the coastal strip immediately around Athens center. In fact, in these municipalities, new forests caused an increase of 22 m^2 per inhabitant (on average) of the indicator in the last decade. Municipalities with decreasing *For* are located in the fringe areas, where in the 1960–2010 period a relevant loss of forest cover occurred (23.6% of the total territory on average). In particular, they are concentrated in the eastern part of Attica (near the new development center of Markopoulo). New forest areas were observed in these municipalities in the last decade.

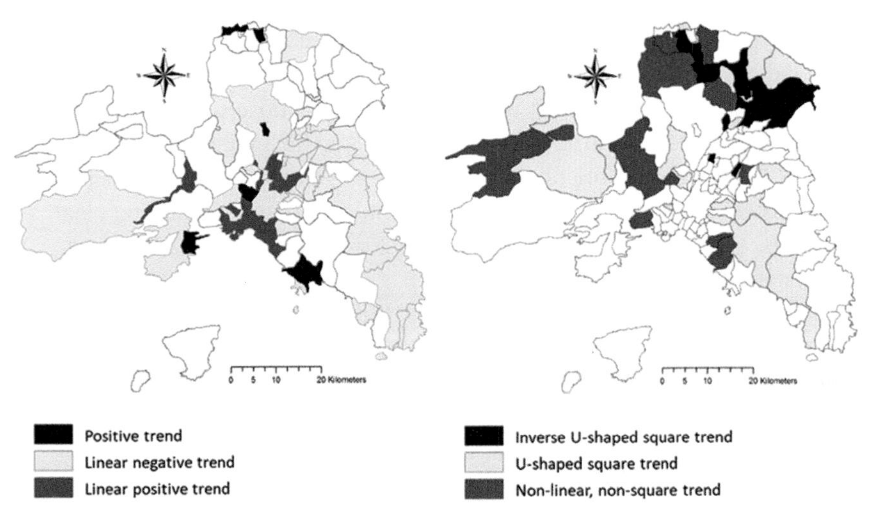

Figure 4.6 Classification of Attica municipalities based on long-term trends in per-capita forest area (left: linear *For* trends; right: non-linear *For* trends).
Source: Authors' elaboration, redrawn from Tomao et al. (2017).

About 31% of the investigated area (26 municipalities) showed a non-linear trend of *For*. Seven of them experienced a relevant increase of *For* in the decade 1960–1970 (+2900 m^2 per capita) followed by a sharp decrease by about 9600 m^2 in 40 years. These municipalities are all located in the northern part of the region, relatively far from the main centers of Athens and Pireo. A total of 19 municipalities (categorized in the U-shaped square trend group) experienced a strong decrease of *For* indicator from 1960–2000 with a reverse trend in the last decade (+65 m^2 per capita). This evidence is probably related to a moderate increase in forest cover. Mixed and heterogeneous class, where neither a linear nor a square trend is identified, includes eight municipalities concentrated in the hilly areas away from the city center. These municipalities are also the larger of the Attica region.

4.4.1 Discussing Socio-Environmental Implications of Forest Expansion and Decline

The present study has analyzed the forest cover trends that occurred during the last decades in a Mediterranean metropolitan area using an original procedure based on the calculation of specific indicators. By profiling the municipalities of Attica, the study has helped to shed light on an urban ecosystem issue with implications on landscape and planning studies. The major evidence of the study is a decrease of forest availability, here measured by forest land per capita, across time in the Attica region, with the only exception of the last decade. The fringe municipalities, *i.e.,* those constituting the 'green belt' clearly recognizable in 1960, experienced the greater reduction of the indicator across time. This phenomenon can be mainly associated with a decrease in forest cover due to the interplay of multiple causes including clear cuts, wildfires, and urban sprawl (Salvati et al., 2012; Kazemzadeh-Zow et al., 2017; Tomao et al., 2021).

Urban sprawl is probably the main driver. Indeed, an uncontrolled urban expansion caused dispersed growth along the seacoast and on Athens' fringe since 1960s (Cecchini et al., 2019; Nickayin et al., 2020). Besides the planned urban expansion aimed at increasing tourism, second-home expansion and decentralization of business activities, many areas were built up informally. This continuous process has reduced the environmental quality and negatively affected the landscape of these regions (Frenkel, 2004; Salvati et al., 2014a; Kosmas et al., 2016; Salvati et al., 2016b). An example, in this sense, is the set of municipalities that gravitate around the new urban expansion center of Messoghia, which developed in a de-regulated way since World War II. Due to this process, a polarization between highly urbanized areas with little forest cover and few municipalities characterized by wider forest cover and then a good efficiency can be found in the last decades. After a first period (1960, 1970) when urban areas with mixed economic functions and high settlement compactness were clearly differenced from rural areas with high forest cover per capita, availability of forest land per capita declined (1980, 1990) along fringe areas surrounding Athens. The important changes observed at the regional and local levels denote the polarization in areas with low and high efficiency in land use along the urban-rural gradient (Zomeni et al., 2008).

The enhanced availability to people of forests in some municipalities located both in densely urbanized areas and in the rural context can be explained with the increase of forest cover. However, the new forests are not due to the establishment of reforestations or afforestation: up to 2000 only 2.4 hectares of new forests were planted (Arabatzis, 2005). Instead, the new forest areas derive from the colonization of marginal and abandoned land by shrubland first and by forests in a second time (Figure 4.7). This phenomenon has been described in several countries in the Mediterranean basin (e.g., Cimini et al., 2013; Barbati et al., 2013; Martín-Forés et al., 2020; Palmero-Iniesta et al., 2021) and is mainly due to the marginalization of rural lands after a long period of decline. Indeed, where rural activities are no longer profitable, farmers abandon agricultural lands that return to the forest

Figure 4.7 Welcome to green city, Acharnes, Northern Athens (Source: authors' photographic archive).

(e.g., Zitti et al., 2015). This is also the case of the Attica region, where the formation of new forests is occurring in rural areas recently abandoned during the economic crisis (Salvati, 2016; Salvati et al., 2016c).

The phenomenon is not only located in rural areas since can also occur within brownfields (Salvati et al., 2014). The increase in forest cover has a multifold positive effect since it enhances the provision of several ecosystem services including, among others, connectivity among natural ecosystems in a green infrastructure perspective (Barbati et al., 2013). However, these newly established forests are among the more fire-prone forest types (Koutsias et al., 2010; Moreira et al., 2011). Therefore, an uncontrolled and unplanned change may represent a danger, since fire risk may be enhanced, particularly in the wildland-urban interface (Corona et al., 2015; Mancini et al., 2018; Moreira et al., 2020).

Active management promoting interventions to reduce fuel, improve functionality and quality of new forests and adapt forest stand structure for people use is required (Tomao et al., 2018). Furthermore, the increase of the accessibility to these new forests may support the use of these spontaneous natural stands as urban parks. In this regard, the newly established forests can help to create a balance between urban development and availability of accessible and usable urban green areas with a relevant positive impact on the quality of the lives of the inhabitants (Kabisch and Haase, 2014; Quatrini et al., 2019). However, now ineffective policies of forest protection and urban containment are acting in the Attica region (Nickayin et al., 2020). Natural places were mostly protected when difficult to access (e.g., mountain areas). Rural places at the fringe of Athens were instead destroyed, fragmented or heavily damaged.

Within the municipalities where the availability of forest land per capita decreased or remained stable, the establishment of new patches of mature vegetation in degraded neighborhoods can be considered as a strategic priority. The recovery of degraded ecosystems while ensuring natural regeneration or artificial intervention (seeding or planting) for production purposes leads also to effective protection of soils. In Mediterranean countries, the reforestation of large areas in a restoration perspective has generally consisted in planting one or a few plant species, often conifers. This approach envisaged a long-term restoration with multiple different steps extended in a large time frame: planting seedlings of conifers at high density, thinning of established individuals to targeted final density, the introduction of hardwood seedlings in the understory, and further silvicultural actions to both conifers and hardwoods to achieve a mixed forest. However, actions rarely went beyond the initial steps and in most cases just stopped in the plantations of conifers (mostly pines). As a result, the area covered by conifers, especially in drylands such as those in the Mediterranean Basin, has largely increased during the second half of the 20th century, increasing the risk of disturbances (fires, pest outbreaks) derived from the lack of management of these areas.

To avoid this process, mixed stands of native species may be preferable since they provide also the advantage of being well known and valued by local communities for their timber and non-timber quality and the restored system is likely less vulnerable to outbreaks of insect pests and diseases, thus supporting a long-term provision of ecosystem services (Bauhus et al., 2017; del Río et al., 2017; Steckel et al., 2020). Improving resilience against future fires outbreak and being capable to provide ecosystems services for the decades to come and under the pressure of climate change, requires a strategic design of species selection, their spatial distribution, food chain and interrelation of main species chains in forest systems. Furthermore, making the system resilient and sustainable requires the study of the history of the area in terms of all the factors affecting trends of contraction and expansion of natural land covers, including forests.

According to this analysis, a more balanced availability of forest land per capita may be reached by investing in increasing forest cover around new developing centers and along the coastal strip. In particular, in coastal areas, forest stands are highly appreciated by people for their role in mitigating microclimate and reducing physical and psychological stress (Tomao et al., 2018) and therefore can support economic activities such as tourism.

4.5 Concluding Remarks

The present chapter analyzes trends over time and spatial patterns of an indicator of efficiency in sustainable governance (per-capita forest land) in Attica proposing an original methodology using specific indicators and statistical approaches. The exploratory analysis was carried out in a metropolitan region characterized – as other cities of southern Europe – by a dispersed settlement expansion and a progressive loss of forests and agricultural land.

Our research has contributed to understanding the forest cover trends in a typical example of a Mediterranean city. In particular, the study has confirmed how rapidly expanding areas such as Attica can be very sensitive to deforestation, particularly due to land take and urbanization of rural peripheries. In this regard, specific measures such as the protection of forest and rural areas are still needed, especially in the rural fringe in order to reduce and/or reverse the processes of degradation of land. Indeed, these measures have been applied only in remote and marginal locations and leaving natural areas surrounding the urban areas of Athens to the negative impacts of infrastructural development.

A reversing trend has been also showing for some municipalities, where an increase of forest cover per capita has been observed. However, this phenomenon is mostly due to forest recolonization after land abandonment. The contribution of new forests to the green infrastructure and ecosystem services provision is not clear and requires new studies, especially regarding their impact on fire risk.

References

Adenle, A. A., Eckert, S., Adedeji, O. I., Ellison, D., and Speranza, C. I. (2020). Human-induced land degradation dominance in the Nigerian Guinea savannah between 2003–2018. Remote sensing applications: society and environment, 19, 100360.

Arabatzis, G. (2005). European Union, Common Agricultural Policy (CAP) and the afforestation of agricultural land in Greece. *New Medit*, 4(4), 48.

Bajocco, S., Ceccarelli, T., Smiraglia, D., Salvati, L., and Ricotta, C. (2016). Modeling the ecological niche of long-term land use changes: The role of biophysical factors. *Ecological Indicators*, 60, 231–236.

Bajocco, S., De Angelis, A., and Salvati, L. (2012). A satellite-based green index as a proxy for vegetation cover quality in a Mediterranean region. *Ecological Indicators*, 23, 578–587.

Bajocco, S., Salvati, L., and Ricotta, C. (2011). Land degradation versus fire: A spiral process? *Progress in Physical Geography*, 35(1), 3–18.

Barbati, A., Corona, P., Salvati, L., and Gasparella, L. (2013). Natural forest expansion into suburban countryside: Gained ground for a green infrastructure?. *Urban Forestry & Urban Greening*, 12(1), 36–43.

Basso, F., Bove, E., Dumontet, S., Ferrara, A., Pisante, M., Quaranta, G., and Taberner, M. (2000). Evaluating environmental sensitivity at the basin scale through the use of geographic information systems and remotely sensed data: an example covering the Agri basin (Southern Italy). *Catena*, 40(1), 19–35.

Bauhus, J., Forrester, D. I., Pretzsch, H., Felton, A., Pyttel, P., and Benneter, A. (2017). Silvicultural options for mixed-species stands. In Mixed-Species Forests (pp. 433–501). Springer, Berlin, Heidelberg.

Blum, W. E. (2013). Soil and land resources for agricultural production: general trends and future scenarios-a worldwide perspective. *International Soil and Water Conservation Research*, 1(3), 1–14.

Breg Valjavec, M., Zorn, M., and Čarni, A. (2018). Human-induced land degradation and biodiversity of Classical Karst landscape: On the example of enclosed karst depressions (dolines). *Land Degradation & Development*, 29(10), 3823–3835.

Carlucci, M., Chelli, F. M., and Salvati, L. (2018). Toward a new cycle: Short-term population dynamics, gentrification, and re-urbanization of Milan (Italy). *Sustainability*, 10(9), 3014.

Carrus, G., Scopelliti, M., Lafortezza, R., Colangelo, G., Ferrini, F., Salbitano, F., Agrimi, M., Portoghesi, L., Semenzato, P. and Sanesi, G. (2015). Go greener, feel better? The positive effects of biodiversity on the well-being of individuals visiting urban and peri-urban green areas. *Landscape and Urban Planning*, 134, 221–228.

Catalán, B., Saurí, D., and Serra, P. (2008). Urban sprawl in the Mediterranean?: Patterns of growth and change in the Barcelona Metropolitan Region 1993–2000. *Landscape and Urban Planning*, 85(3–4), 174–184.

Cecchini, M., Zambon, I., Pontrandolfi, A., Turco, R., Colantoni, A., Mavrakis, A., and Salvati, L. (2019). Urban sprawl and the 'olive' landscape: Sustainable land management for 'crisis' cities. *GeoJournal*, 84(1), 237–255.

Chas-Amil, M. L., Touza, J., and García-Martínez, E. (2013). Forest fires in the wildland–urban interface: a spatial analysis of forest fragmentation and human impacts. *Applied Geography*, 43, 127–137.

Chorianopoulos, I., Pagonis, T., Koukoulas, S., and Drymoniti, S. (2010). Planning, competitiveness and sprawl in the Mediterranean city: *The Case of Athens*. Cities, 27(4), 249–259.

Chorianopoulos, I., Tsilimigkas, G., Koukoulas, S., and Balatsos, T. (2014). The shift to competitiveness and a new phase of sprawl in the Mediterranean city: Enterprises guiding growth in Messoghia–Athens. Cities, 39, 133–143.

Cimini, D., Tomao, A., Mattioli, W., Barbati, A., and Corona, P., (2013). Assessing impact of forest cover change dynamics on high nature value farmland in Mediterranean mountain landscape. *Annals of Silvicultural Research* 37(1), 29–37.

Ciommi, M., Chelli, F.M., Carlucci, M., and Salvati, L. (2018). Urban growth and demographic dynamics in southern Europe: Toward a new statistical approach to regional science. *Sustainability* (Switzerland), 10(8), 2765.

Ciommi, M., Chelli, F.M., and Salvati, L. (2019). Integrating parametric and non-parametric multivariate analysis of urban growth and commuting patterns in a European metropolitan area. *Quality and Quantity*, 53(2), 957–979.

Ciommi, M., Gigliarano, C., Emili, A., Taralli, S., and Chelli, F.M. (2017). A new class of composite indicators for measuring well-being at the local level: An application to the Equitable and Sustainable Well-being (BES) of the Italian Provinces. *Ecological Indicators*, 76, 281–296.

Coluzzi, R., D'Emilio, M., Imbrenda, V., Giorgio, G. A., Lanfredi, M., Macchiato, M., Simoniello, T., and Telesca, V. (2019). Investigating climate variability and long-term vegetation activity across heterogeneous Basilicata agroecosystems. *Geomatics, Natural Hazards and Risk*, 10(1), 168–180.

Corona, P., Ascoli, D., Barbati, A.; Bovio, G., Colangelo, G., Elia, M., Garfì, V., Iovino, F., Lafortezza, R. and Leone, V. (2015). Integrated forest management to prevent wildfires under Mediterranean environments. *Annals of Silvicultural Research*, 39, 1–22

Couch, C., Petschel-Held, G. and Leontidou, L. 2007. Urban Sprawl In Europe: Landscapes, *Land-use Change and Policy*. Blackwell, London.

Curebal, I., Efe, R., Soykan, A., and Sonmez, S. (2015). Impacts of anthropogenic factors on land degradation during the anthropocene in Turkey. *Journal of Environmental Biology*, 36(1), 51.

del Río, M., Pretzsch, H., Ruíz-Peinado, R., Ampoorter, E., Annighöfer, P., Barbeito, I., Bielak, K., Brazaitis, G., Coll, L., Drössler, L., Fabrika, M., Forrester, D.I., Heym, M., Hurt, V., Kurylyak, V., Löf, M., Lombardi, F., Madrickiene, E., Matović, B., Mohren, F., Motta, R., den Ouden, J., Pach, M., Ponette, Q., Schütze, G., Skrzyszewski, J., E. Sramek, V., Sterba, H., Stojanović, D., Svoboda, M., Zlatanov, T.M., and Bravo-Oviedo, A. (2017). Species interactions increase the temporal stability of community productivity in Pinus sylvestris-Fagus sylvatica mixtures across Europe. *Journal of Ecology*, 105, 1032–1043.

Delfanti, L., Colantoni, A., Recanatesi, F., Bencardino, M., Sateriano, A., Zambon, I., and Salvati, L. (2016). Solar plants, environmental degradation and local socioeconomic contexts: A case study in a Mediterranean country. *Environmental Impact Assessment Review*, 61, 88–93.

Di Feliciantonio, C., and Salvati, L. (2015). 'Southern' Alternatives of Urban Diffusion: Investigating Settlement Characteristics and Socio-Economic Patterns in Three Mediterranean Regions. Tijdschrift voor economische en sociale geografie, 106(4), 453–470.

Di Feliciantonio, C., Salvati, L., Sarantakou, E., and Rontos, K. (2018). Class diversification, economic growth and urban sprawl: evidences from a pre-crisis European city. *Quality & Quantity*, 52(4), 1501–1522.

Dobbs, C., Kendal, D., and Nitschke, C.R. 2014. Multiple ecosystem services and disservices of the urban forest establishing their connections with landscape structure and sociodemographics. *Ecological Indicators*, 43, 44–55.

Duvernoy, I., Zambon, I., Sateriano, A., and Salvati, L. (2018). Pictures from the other side of the fringe: Urban growth and peri-urban agriculture in a post-industrial city (Toulouse, France). *Journal of Rural Studies*, 57, 25–35.

Escobedo, F. J., Giannico, V., Jim, C. Y., Sanesi, G., and Lafortezza, R. (2019). Urban forests, ecosystem services, green infrastructure and nature-based solutions: Nexus or evolving metaphors? *Urban Forestry & Urban Greening*, 37, 3–12.

European Environment Agency (EEA) 2006. Urban sprawl in Europe - The ignored challenge. *European Environmental Agency*, Report no. 10, Copenhagen.

European Environment Agency (EEA) 2010. Mapping guide for a European Urban Atlas. *European Environment Agency*, Copenhagen.

Ferrara, A., Kosmas, C., Salvati, L., Padula, A., Mancino, G., and Nolè, A. (2020). Updating the MEDALUS-ESA Framework for Worldwide Land Degradation and Desertification Assessment. *Land Degradation & Development*, 31(12), 1593–1607.

Frenkel, A. (2004). The potential effect of national growth-management policy on urban sprawl and the depletion of open spaces and farmland. *Land Use Policy*, 21, 357–369.

Hammad, A., and Tumeizi, A. (2012). Land degradation: socioeconomic and environmental causes and consequences in the eastern Mediterranean. *Land Degradation & Development*, 23(3), 216–226.

Hasse, J. E., and Lathrop, R. G. (2003). Land resource impact indicators of urban sprawl. *Applied Geography*, 23(2–3), 159–175.

Imbrenda, V., Coluzzi, R., Lanfredi, M., Loperte, A., Satriani, A., and Simoniello, T. (2018). Analysis of landscape evolution in a vulnerable coastal area under natural and human pressure. *Geomatics, Natural Hazards and Risk*, 9(1), 1249–1279.

Imbrenda, V., D'Emilio, M., Lanfredi, M., Ragosta, M., and Simoniello, T. (2013). Indicators of Land Degradation Vulnerability Due to Anthropic

Factors: Tools for an Efficient Planning Available online: www.igi-global.com/chapter/indicators-of-land-degradation-vulnerability-due-to-anthropic-factors/95002

Kabisch, N., and Haase, D. (2014). Green justice or just green? Provision of urban green spaces in Berlin, Germany. *Landscape and Urban Planning*, 122, 129–139.

Kairis, O., Karavitis, C., Kounalaki, A., Salvati, L., and Kosmas, C. (2013). The effect of land management practices on soil erosion and land desertification in an olive grove. *Soil Use and Management*, 29(4), 597–606.

Kairis, O., Karavitis, C., Salvati, L., Kounalaki, A., and Kosmas, K. (2015). Exploring the impact of overgrazing on soil erosion and land degradation in a dry Mediterranean agro-forest landscape (Crete, Greece). *Arid Land Research and Management*, 29(3), 360–374.

Karamesouti, M., Detsis, V., Kounalaki, A., Vasiliou, P., Salvati, L., and Kosmas, C. (2015). Land-use and land degradation processes affecting soil resources: Evidence from a traditional Mediterranean cropland (Greece). *Catena*, 132, 45–55.

Kazemzadeh-Zow, A., Zanganeh Shahraki, S., Salvati, L., and Samani, N. N. (2017). A spatial zoning approach to calibrate and validate urban growth models. *International Journal of Geographical Information Science*, 31(4), 763–782.

Kelly, C., Ferrara, A., Wilson, G. A., Ripullone, F., Nolè, A., Harmer, N., and Salvati, L. (2015). Community resilience and land degradation in forest and shrubland socio-ecological systems: Evidence from Gorgoglione, Basilicata, Italy. *Land Use Policy*, 46, 11–20.

Konijnendijk, C. C., Nilsson, K., Randrup, T. B., Schipperijn, J. (2005). Urban forests and trees: a reference book. *Springer Science & Business Media.*

Kosmas, C., Kairis, O., Karavitis, C., Ritsema, C., Salvati, L., Acikalin, S., ... and Ziogas, A. (2014). Evaluation and selection of indicators for land degradation and desertification monitoring: methodological approach. *Environmental Management*, 54(5), 951–970.

Kosmas, C., Karamesouti, M., Kounalaki, K., Detsis, V., Vassiliou, P., and Salvati, L. (2016). Land degradation and long-term changes in agro-pastoral systems: An empirical analysis of ecological resilience in Asteroussia-Crete (Greece). *Catena*, 147, 196–204.

Koutsias, N., Martínez-Fernández, J., and Allgöwer, B. (2010). Do factors causing wildfires vary in space? Evidence from geographically weighted regression. *GIScience & Remote Sensing*, 47(2), 221–240.

Mancini, L. D., Elia, M., Barbati, A., Salvati, L., Corona, P., Lafortezza, R., and Sanesi, G. (2018). Are wildfires knocking on the built-up areas door? *Forests*, 9(5), 234.

Markwell, N. and Gladwin, T. E. (2020). Shinrin-yoku (forest bathing) reduces stress and increases people's positive affect and well-being in comparison with its digital counterpart. *Ecopsychology*, 12(4), 247–256.

Martín-Forés, I., Magro, S., Bravo-Oviedo, A., Alfaro-Sánchez, R., Espelta, J. M., Frei, T., Valdés-Correcher, E., Rodríguez Fernández-Blanco, C., Winkel, G., Gerzabek, G., González-Martínez, S. C., Hampe, A., and Valladares, F. (2020) Spontaneous forest regrowth in South-West Europe: Consequences for nature's contributions to people. *People and Nature*, https://doi.org/10.1002/pan3.10161

Masini, E., Tomao, A., Barbati, A., Corona, P., Serra, P., and Salvati, L. (2019). Urban growth, land-use efficiency and local socioeconomic context: a comparative analysis of 417 metropolitan regions in Europe. *Environmental Management*, 63(3), 322–337.

Moreira, F., Viedma, O., Arianoutsou, M., Curt, T., Koutsias, N., Rigolot, F., Barbati, A., Corona, P., Vaz, P., Xanthopoulos, G., Mouillot, F. And Bilgili, E. (2011). Landscape wildfire interactions in southern Europe: implications for landscape management. *Journal of Environmental Management*, 92, 2389–2402.

Moreira, F., Ascoli, D., Safford, H., Adams, M. A., Moreno, J. M., Pereira, J. M. C., Catry, F. X., Armesto, J., Bond, W., Gonzalez, M., Koutsias, N., McCar, L., Price, O., Pausas, J. G., Rigolot, E., Stephens, S., Tavsanoglu, C., Vallejo, V. R., van Wilgen, B. W., Xanthopoulos, G., and Fernandes, P. M. (2020). Wildfire management in Mediterranean-type regions: paradigm change needed, *Environmental Research Letters*, 15, 011001, https://doi.org/10.1088/1748-9326/ab541e

Nickayin, S. S., Tomao, A., Quaranta, G., Salvati, L., and Gimenez Morera, A. (2020). Going toward Resilience? Town Planning, Peri-Urban Landscapes, and the Expansion of Athens, Greece. *Sustainability*, 12(24), 10471.

Palmero-Iniesta, M., Pino, J., Pesquer, L., and Espelta, J. M. (2021). Recent forest area increase in Europe: expanding and regenerating forests differ in their regional patterns, drivers and productivity trends. European *Journal of Forest Research*, 1–13.

Perdigao, V., and Christensen, S. (2000). The LACOAST atlas: Land cover changes in European coastal zones. SPI 00.39 EN, European Commission, DJ-Joint Research Centre, Ispra.

Pili, S., Grigoriadis, E., Carlucci, M., Clemente, M., and Salvati, L. (2017). Towards sustainable growth? A multi-criteria assessment of (changing) urban forms. *Ecological Indicators*, 76, 71–80.

Quaranta, G., Salvia, R., Salvati, L., Paola, V. D., Coluzzi, R., Imbrenda, V., and Simoniello, T. (2020). Long-term impacts of grazing management on land degradation in a rural community of Southern Italy: Depopulation matters. *Land Degradation & Development*, 31(16), 2379–2394.

Quatrini, V., Tomao, A., Corona, P., Ferrari, B., Masini, E., and Agrimi, M. (2019). Is new always better than old? Accessibility and usability of the urban green areas of the municipality of Rome. *Urban Forestry & Urban Greening*, 37, 126–134.

Radeloff, V. C., Hammer, R. B., Stewart, S. I., Fried, J. S., Holcomb, S. S., and McKeefry, J. F. (2005). The wildland–urban interface in the United States. *Ecological Applications*, 15(3), 799–805.

Recanatesi, F., Clemente, M., Grigoriadis, E., Ranalli, F., Zitti, M., and Salvati, L. (2016). A fifty-year sustainability assessment of Italian agro-forest districts. *Sustainability*, 8(1), 32.

Reed, M.S., Buenemann, M., Atlhopheng, J., Akhtar-Schuster, M., Bachmann, F., Bastin, G., Bigas, H., Chanda, R., Dougill, A.J. and Essahli, W. (2011). Cross-scale monitoring and assessment of land degradation and sustainable land management: A methodological framework for knowledge management. *Land Degradation and Development* 22, 261–271.

Reyes-Riveros, R., Altamirano, A., De La Barrera, F., Rozas, D., Vieli, L., and Meli, P. (2021). Linking public urban green spaces and human well-being: A systematic review. Urban Forestry & Urban Greening, 127105.

Salvati, L. (2014). Exurban Development and Landscape Diversification in a Mediterranean Peri-urban Area. *Scottish Geographical Journal*, 130(1), 22–34.

Salvati, L. (2016). The Dark Side of the Crisis: Disparities in per Capita income (2000–12) and the Urban-Rural Gradient in Greece. Tijdschrift voor economische en sociale geografie, 107(5), 628–641.

Salvati, L. and Zitti, M. (2005). Land degradation in the Mediterranean Basin: Linking bio-physical and economic factors into an ecological perspective. Biota, 6, 67–77.

Salvati, L. and Zitti, M. (2009). The environmental 'risky' region: identifying land degradation processes through integration of socio-economic and ecological indicators in a multivariate regionalization model. *Environmental Management*, 44(5), 888.

Salvati, L., Bajocco, S., Ceccarelli, T., Zitti, M., and Perini, L. (2011). Towards a process-based evaluation of land vulnerability to soil degradation in Italy. *Ecological Indicators*, 11(5), 1216–1227.

Salvati, L., Ferrara, C., and Ranalli, F. (2014). Changes at the fringe: Soil quality and environmental vulnerability during intense urban expansion. *Eurasian Soil Science*, 47(10), 1069–1075.

Salvati, L., Perini, L., Sabbi, A., and Bajocco, S. (2012). Climate Aridity and Land Use Changes: A Regional-Scale Analysis. *Geographical Research* 50(2), 193–203.

Salvati, L., Quatrini, V., Barbati, A., Tomao, A., Mavrakis, A., Serra, P., Sabbi, A., Merlini, P., and Corona, P. (2016a). Soil occupation efficiency and landscape conservation in four Mediterranean urban regions. Urban *Forestry & Urban Greening*, 20, 419–427.

Salvati, L., Ridolfi, E., Pujol, D. S., and Ruiz, P. S. (2016b). Latent sprawl, divided Mediterranean landscapes: Urban growth, swimming pools, and the socio-spatial structure of Athens, Greece. *Urban Geography*, 37(2), 296–312.

Salvati, L., Sateriano, A. and Bajocco, S. (2013). To grow or to sprawl? Evolving land cover relationships in a compact Mediterranean city region. Cities 30, 113–121.

Salvati, L., Sateriano, A., and Grigoriadis, E. (2016c). Crisis and the city: profiling urban growth under economic expansion and stagnation. *Letters in Spatial and Resource Sciences*, 9(3), 329–342.

Salvati, L. and Serra, P. (2016). Estimating rapidity of change in complex urban systems: A multidimensional, local-scale approach. *Geographical Analysis*, 48(2), 132–156.

Simoniello, T., Coluzzi, R., Imbrenda, V., and Lanfredi, M. (2015). Land cover changes and forest landscape evolution (1985–2009) in a typical Mediterranean agroforestry system (high Agri Valley). *Natural Hazards and Earth System Sciences*, 15(6), 1201–1214.

Smiraglia, D., Ceccarelli, T., Bajocco, S., Salvati, L., and Perini, L. (2016). Linking trajectories of land change, land degradation processes and ecosystem services. *Environmental Research*, 147, 590–600.

Smiraglia, D., Tombolini, I., Canfora, L., Bajocco, S., Perini, L., and Salvati, L. (2019). The Latent Relationship Between Soil Vulnerability to Degradation and Land Fragmentation: A Statistical Analysis of Landscape Metrics in Italy, 1960–2010. *Environmental Management*, 64(2), 154–165.

Steckel, M., del Río, M., Heym, M., Aldea, J., Bielak, K., Brazaitis, G., ... and Pretzsch, H. (2020). Species mixing reduces drought susceptibility of

Scots pine (Pinus sylvestris L.) and oak (Quercus robur L., Quercus petraea (Matt.) Liebl.)–Site water supply and fertility modify the mixing effect. *Forest Ecology and Management*, 461, 117908.

Tomao, A., Quaranta, G., Salvia, R., Vinci, S., & Salvati, L. (2021). Revisiting the 'southern mood'? Post-crisis Mediterranean urbanities between economic downturns and land-use change. Land Use Policy, 111, 105740.

Tomao, A., Quatrini, V., Corona, P., Ferrara, A., Lafortezza, R., and Salvati, L. (2017). Resilient landscapes in Mediterranean urban areas: Understanding factors influencing forest trends. *Environmental research*, 156, 1–9.

Tomao, A., Secondi, L., Carrus, G., Corona, P., Portoghesi, L., and Agrimi, M. (2018). Restorative urban forests: Exploring the relationships between forest stand structure, perceived restorativeness and benefits gained by visitors to coastal Pinus pinea forests. *Ecological Indicators*, 90, 594–605.

Tomao, A., Secondi, L., Corona, P., Carrus, G., and Agrimi, M. (2016). Exploring individuals' well-being visiting urban and peri-urban green areas: a quantile regression approach. *Agriculture and Agricultural Science Procedia*, 8, 115–122.

Wilson, G. A. and Juntti, M. (Eds.). (2005). Unravelling desertification: policies and actor networks in Southern Europe. *Wageningen Academic Publishers*.

Winslow, M., Akhtar-Schuster, M., Martius, C., Stringer, L., Thomas, R. and Vogt, J. (2011). Special issue on understanding dryland degradation trends. *Land Degradation and Development*, 22, 145–312.

Xiao, R., Liu, Y., Fei, X., Yu, W., Zhang, Z., and Meng, Q. (2019). Ecosystem health assessment: A comprehensive and detailed analysis of the case study in coastal metropolitan region, eastern China. *Ecological Indicators*, 98, 363–376.

Xie, H., Zhang, Y., Wu, Z., and Lv, T. (2020). A bibliometric analysis on land degradation: Current status, development, and future directions. Land, 9(1), 28.

Zambon, I., Benedetti, A., Ferrara, C., and Salvati, L. (2018). Soil matters? A multivariate analysis of socioeconomic constraints to urban expansion in Mediterranean Europe. *Ecological Economics*, 146, 173–183.

Zambon, I., Colantoni, A., Carlucci, M., Morrow, N., Sateriano, A., and Salvati, L. (2017). Land quality, sustainable development and environmental degradation in agricultural districts: A computational approach based on entropy indexes. *Environmental Impact Assessment Review*, 64, 37–46.

Zipperer, W. C., Foresman, T. W., Walker, S. P., and Daniel, C. T. (2012). Ecological consequences of fragmentation and deforestation in an urban landscape: a case study. *Urban Ecosystems*, 15(3), 533–544.

Zitti, M., Ferrara, C., Perini, L., Carlucci, M., and Salvati, L. (2015). Long-term urban growth and land use efficiency in Southern Europe: Implications for sustainable land management. *Sustainability*, 7(3), 3359–3385.

Zomeni, M., Tzanopoulos, J., and Pantis, J. D. (2008). Historical analysis of landscape change using remote sensing techniques: An explanatory tool for agricultural transformation in Greek rural areas. *Landscape and Urban Planning*, 86(1), 38–46.

5

Welcome to 'Forestscapes': In Between Urban Reality and Rural Idyll

Matteo Clemente[1], Adriano Conte[2], Giovanni Quaranta[3], and Luca Salvati[4]

[1]Department of Architecture and Project, 'Sapienza' University of Rome, Via Flaminia 359, I-00196 Rome, Italy
[2]Italian Council for Agricultural Research and Economics (CREA), Research Centre for Forestry and Wood, Via Valle della Quistione 27, I-00166 Rome, Italy
[3]Department of Mathematics, Computer Science and Economics Department, University of Basilicata, Viale dell'Ateneo Lucano, I-85100 Potenza, Italy
[4]Department of Methods and Models for Economics, Territory and Finance (MEMOTEF), Faculty of Economics, Sapienza University of Rome, Via del Castro Laurenziano 9, I-00161 Rome, Italy
E-mail: matteo.clemente@uniroma1.it; adriano.conte@crea.gov.it; giovanni.quaranta@unibas.it; luca.salvati@uniroma1.it

Abstract

Landscape is a complex system of interrelated biophysical and anthropogenic elements that changes rapidly according to external pressures. This system definitely needs permanent monitoring systems to inform effective conservation policies. Landscape analysis benefits of integrated approaches encompassing geography, environmental science and information systems to evaluate structure, form, diversity and dynamics of different land-use types. While nowadays monitoring techniques for land-use changes detection are common and well defined, the evaluation of specific landscape elements over time is a research issue that deserves further efforts in both the theoretical and practical perspective. The Mediterranean rural landscape is one of the most rich in term of cultural and natural biodiversity in the world since it is characterized by specific structures and multifaceted functions. This landscape type, however, is threatened by drastic land-use changes in the area mainly driven by urbanization, infrastructural development and cropland abandonment.

Keywords: City, Suburbanization, Architecture, Rural areas, Local communities, Southern Europe.

5.1 Introduction

The landscape is a complex system of interrelated biophysical and anthropogenic elements that changes rapidly according to external pressures. This system definitely needs permanent monitoring systems to inform effective conservation policies. Landscape analysis benefits of integrated approaches encompassing geography, environmental science and information systems to evaluate structure, form, diversity and dynamics of different land-use types. While monitoring techniques for land-use change detection seem to be rather well defined nowadays, the evaluation of specific landscape elements over time is a research issue that deserves further efforts in both the theoretical and practical perspectives. The Mediterranean rural landscape is one of the richest in terms of cultural and natural biodiversity in the world since it is characterized by specific structures and multifaceted functions. This landscape type, however, is threatened by drastic land-use changes in the area mainly driven by urbanization, infrastructural development and cropland abandonment.

Responding to increasing processes of globalization, urbanization and socio-economic change, the world has moved towards an urban dimension. At the same time, cities – whether large or small, including neighbourhoods, city-centers, suburban or peri-urban areas – are constantly changing (UN-Habitat, 2012). They are built, rebuilt, transformed and used by people for different functions and grow in complex ways in terms of population, size, socioeconomic patterns, geopolitical settings (Hall, 1998; Batty and Marshall, 2012) and over-consumptive use of natural resources. Considered as the most complex, dynamic and never finished artifacts created by human activity (UN-Habitat, 2010; Zamenopoulos and Alexiou, 2012; Portugali et al., 2012) they reproduce the interactions among biophysical, social, environmental and economic processes at the local, regional and global scales (Kötter, 2004; Swyngedouw and Heynen, 2003).

According to Bryant et al. (1982), Antrop (2000), Aguilar (2008) and Gargiulo Morelli and Salvati (2010), Land Cover Change (LCCs) are seen as the result of a complex socio-economic system involving several interacting agents. Therefore, large Mediterranean urban regions can be regarded as laboratories of LCCs where a multifaceted stratification of immediate and underlying causes determines land conversion (Soliman et al., 2004; Salvati and Zitti (2007); Jomaa et al., 2008; Marull et al., 2009). The informal action of the private agents coupled with a deregulated planning system promoted the spontaneous 'compact growth' until the 1980s and the subsequent, dispersed expansion that was called 'sprawl' (e.g. Couch et al., 2007) (Figure 5.1).

Due to the stratification of the socio-economic forces impacting on the landscape, the classical models interpreting LCCs and the relationships among land cover types are hardly applicable to the Mediterranean basin (see Briassoulis, 2001; Garcia Latorre et al., 2001; Antrop, 2004; Polyzos et al., 2008; Geri et al., 2010, for reviews). The socio-economic context is complicated by local forces and constrained by important biophysical factors (e.g. land availability) (Leontidou, 1990). In this view, we refer to Coccossis

Figure 5.1 Old (compact) morphologies and new dispersed settlements in the metropolitan nebulosa: Athens from Penteli woodland at dusk.

Source: Authors' photographic archive.

(1991) which proposes the notion of a dynamic 'ecological equilibrium' in land cover transitions as the result of changes in the equilibrium among several factors including population, resources, technology and institutions. This is even truer considering that several Mediterranean urban regions are encompassing (and often deforming) the boundaries of the administrative domains which contain the main agglomeration (Kasanko et al., 2008).

The present chapter argued that landscape changes resulting from the complex political, economical, social, cultural and ecological processes that form certain types of urban contexts in Mediterranean cities are permeating the intimate relationship between environment and urban form. Processes of urban expansion and economic restructuring at the city-region scale are interconnected and produce a particular socio-environment metabolism that materializes through new urban morphologies (Figure 5.1). One typical example is provided by the traditional rural landscapes along the coasts or around the main cities in southern Europe (Couch et al., 2007). These landscapes are characterized by diversity in land-use and land cover, high vegetation and animal biodiversity, traditional crops and agricultural practices (irrigation, mechanization, rotation), local communities, archaeological places and millenary cultures which are being progressively threatened by urban sprawl (Galli et al., 2010). A framework for permanent monitoring and sustainable planning of these Mediterranean landscapes is particularly needed in the current socio-economic context. Our study provides a preliminary contribution to this interesting and complex research theme.

5.2 Land–Use Change in Southern Europe

Since the 1980s, Mediterranean cities underwent a rapid transition from the traditional 'compact' model to various steps of a more 'dispersed' form characterized by huge expansion around the urban area (Schneider and Woodcock, 2008). In the strictly 'compact' phase, urbanization primarily involved poor-quality lands, such as pastures, abandoned fields and areas with poor spontaneous vegetation like maquis and shrublands. During this phase, a few agricultural and forest land was converted to built-up settlements by clearcutting and fire or indirectly by land fragmentation due to infrastructure development (Genske, 2003; Portnov and Safriel, 2004; Johnson and Lewis, 2007).

In the most recent 'sprawling' phase, an impressively growing surface of rural lands underwent LCCs (Muñoz, 2003; Catalàn et al., 2008; Nikolakopoulos et al., 2005; Weber et al., 2005; Cakir et al., 2008; Gospodini,

2009). Dispersed villages and clusters of isolated buildings, commercial units, industrial settlements and tourism resorts, all linked by dense road networks, are becoming the ubiquitous elements of the Mediterranean suburban landscape (Chorianopoulos et al., 2010). However, the increasing proportion of rural high-quality land experienced changes in use and cover due to dispersed urbanization with important implications on landscape structure, quality and diversity and the progressive destruction of traditional landscape elements (Weber et al., 2005; Catalàn et al., 2008; Frondoni et al., 2011).

Cultural landscapes in rural areas represent the 'combined works of nature and man' (Birks et al., 1998; David et al., 1998; Farina, 2000; Fowler, 2003) and are widely regarded as coupled natural-human systems due to the emergence of traditional techniques and practices that have shaped the land for centuries (Head, 2000; Marcucci, 2000; Agnoletti, 2007; Green, 2012). Many cultural landscapes are currently listed in the World Cultural Heritage list of UNESCO since they are 'irreplaceable sources of life and inspiration' (UNESCO, 2012). However, human activities in rural and peri-urban areas can both create cultural landscapes with high aesthetic, cultural and ecological value or may result in land degradation and loss of biodiversity (Droste et al., 1995; Fraser Hart, 1998; Garcia Latorre et al., 2001; Wrbka et al., 2004).

5.3 Land–Use Change at The 'Forestscape': The Case of Attica

Attica, extending more than 3,000 km^2, is a Greek region subdivided into 115 municipalities of which 58 formed the strictly urban area (430 km^2), consisting mostly of mountains bordering the urban area of Athens that

Table 5.1 Population and building indicators in Attica, 1961–2011

Year	Population Density (*)	Annual Population Growth (%)	Suburban Population (%)	People Per Building (**)
1961	669	2.5	8.4	14.9
1971	914	2.6	8.2	9.5
1981	1103	1.8	9.2	7.2
1991	1154	0.4	12.4	5.7
2001	1231	0.6	15.1	5.2
2011	1248	0.3	15.6	5.0

(*) Inhabitants per km^2; (**) includes residential and service buildings; (***) provisional figures from ESYE population census and building registers.
Source: Authors' own elaboration on ESYE-ELSTAT data.

occupies a relatively flat territory. Three coastal plains are located in the Attica region outside the urban area: the Messoghia plain, the Marathon plain and the Thriasio plain. While population density increased rapidly from 0.7 thousand inhabitants per km^2 in the early 1960s to 1.25 thousand inhabitants per km^2 in the early 2010s, the annual population growth rate declined progressively averaging 0.3% in the most recent decade.

The ratio of suburban population to the total resident population in the area continuously increased in contrast with 'people per building' indicator that showed a three-fold decrease. Important changes in the urban and peri-urban landscape were observed in the last fifty years in Athens. In 1960, the landscape was typically mixed with a predominance of forests and croplands. Urban fabric covered 12% of the investigated surface area, respectively 53% and 5% in urban and the remaining (rural) part of Attica. Urban areas increased during 1960–2009 (0.36% and 0.53% per year in

Figure 5.2 Urban expansion in Athens' fringe (left: 2004; right: 2013). Source: authors' own elaboration.

1960–1990 and 1990–2009, respectively). As expected, in the former phase, urban areas grew more in the strictly urban area than in the rural area while in the latter phase the reverse pattern was observed.

5.4 Urban Morphology and 'Forestscapes'

The increase in the per-capita surface of built-up areas was found higher in the latter than in the former phase. At the city-region level, cropland decreased by

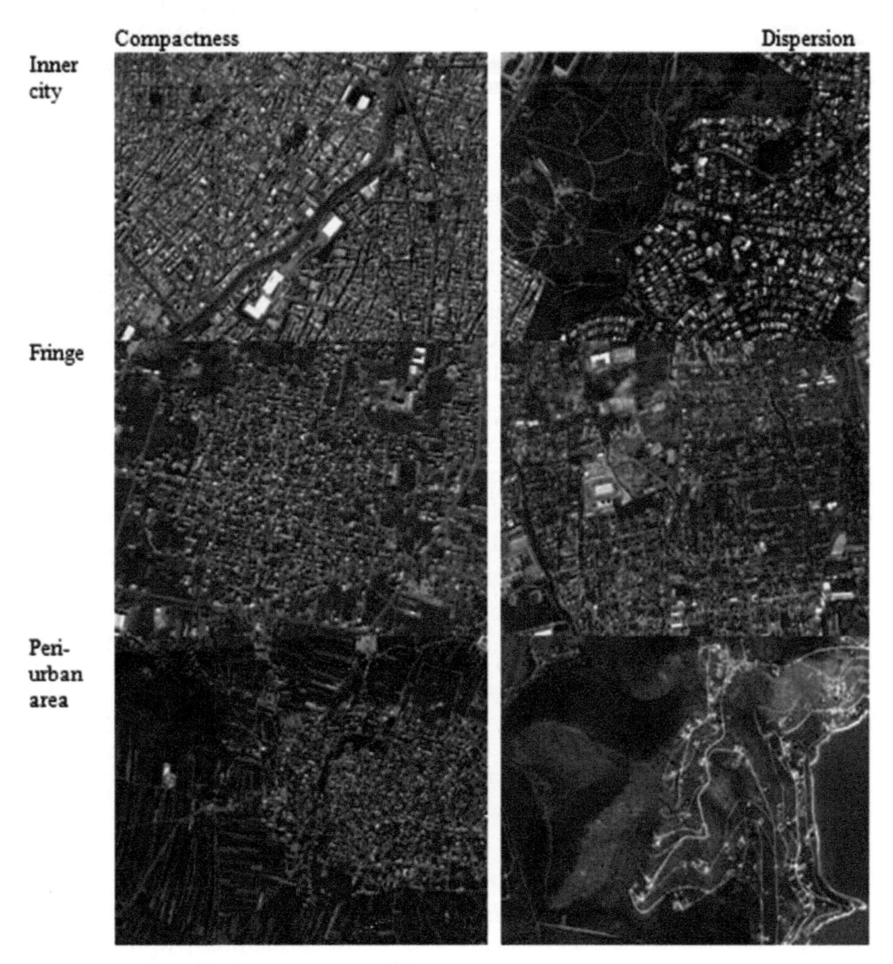

Figure 5.3 Examples of diverging urban forms (from compactness to settlement dispersion) along the urban-to-rural gradient.

Source: Authors' own elaboration.

−0.29% and −0.32% per year in the former and latter phases, respectively). However, in the urban area, cropland experienced an initial decrease followed by a later increase, while in the rural area, cropland decrease was more rapid in the latter than in the former phase. Forest land showed a pattern similar to that observed for cropland at the regional level (Figure 5.2).

Figure 5.3 illustrates different settlement morphologies based on diachronic satellite images covering selected fringe areas. In accordance with the findings illustrated previously, the analysis indicates 1. a diffused expansion of low-density settlements, 2. the formation (or the consolidation) of a mixed and enthropic landscape and 3. habitat fragmentation and important losses in the natural resources' stock.

5.5 Concluding Remarks

Mixed narrative and quantitative approaches based on comparative analysis of relevant cities' casestudies represents a tool for the in-depth understanding of the intimate relationship among contemporary processes of urbanization and landscape changes on the fringe. Original processes of formation and consolidation of a new landscape asset, that we called 'fringscape', should be better investigated. 'Fringscape' is characterized by land-use mixité, patchiness, a fractal design of landscape elements, enthropy and confusion in the spatial distribution of settlements and the disappearence of important rural landmarks. Moreover, future research on the urban change phenomenon should take into account more effectively the processes that constitute urban metabolism since it is crucial to understand the forms and socio-environmental dynamics of cities. Moreover, such a type of analysis, revealing the socio-ecological metabolism of the urban scale, constitutes the basis to explore those flows that constitute an important part of the city metabolism.

References

Antrop, M. (2000). Changing patterns in the urbanized countryside of Western Europe. *Landscape Ecology*, 15(3), 257–270.

Bajocco, S., Ceccarelli, T., Smiraglia, D., Salvati, L., and Ricotta, C. (2016). Modeling the ecological niche of long-term land use changes: The role of biophysical factors. *Ecological Indicators*, 60, 231–236.

Bajocco, S., De Angelis, A., and Salvati, L. (2012). A satellite-based green index as a proxy for vegetation cover quality in a Mediterranean region. *Ecological Indicators*, 23, 578–587.

Bajocco, S., Salvati, L., and Ricotta, C. (2011). Land degradation versus fire: A spiral process? *Progress in Physical Geography*, 35(1), 3–18.

Brouwer, F. B., Thomas, A. J., and Chadwick, M. J. (1991). Land use changes in Europe. Processes of change, environmental transformations and future patterns. Kluwer Academic Publishers, Dordrecht, 530 pp.

Bryant, C. R., Russwurm, L. H., and McLellan, A. G. (1982). The city's countryside. Land and its management in the rural-urban fringe. Longman, London and New York, 249 pp.

Çakir, G., Ün, C., Baskent, E. Z., Köse, S., Sivrikaya, F., and Keleş, S. (2008). Evaluating urbanization, fragmentation and land use/land cover change pattern in Istanbul city, Turkey from 1971 to 2002. *Land Degradation & Development*, 19(6), 663–675.

Carlucci, M., Chelli, F. M., and Salvati, L. (2018). Toward a new cycle: Short-term population dynamics, gentrification, and re-urbanization of Milan (Italy). *Sustainability* (Switzerland), 10(9), 3014.

Catalàn, B., Sauri, D., and Serra, P. (2008). Urban sprawl in the Mediterranean? Patterns of growth and change in the Barcelona Metropolitan Region 1993–2000. *Landscape and Urban Planning*, 85, 174–184.

Cecchini, M., Zambon, I., Pontrandolfi, A., Turco, R., Colantoni, A., Mavrakis, A., and Salvati, L. (2019). Urban sprawl and the 'olive' landscape: Sustainable land management for 'crisis' cities. *GeoJournal*, 84(1), 237–255.

Chorianopoulos, I., Pagonis, T., Koukoulas, S., and Drymoniti, S. (2010). Planning, competitiveness and sprawl in the Mediterranean city: The case of Athens. Cities 27, 249–259.

Christopoulou, O., Polyzos, S., and Minetos, D. (2007). Peri-urban and urban forests in Greece: obstacle or advantage to urban development? *Journal of Environmental Management* 18, 382–395.

Ciommi, M., Chelli, F. M., Carlucci, M., and Salvati, L. (2018). Urban growth and demographic dynamics in southern Europe: Toward a new statistical approach to regional science. *Sustainability* (Switzerland), 10(8), 2765.

Couch, C., Petschel-held, G., and Leontidou, L. (2007). Urban Sprawl In Europe: Landscapes, *Land-use Change and Policy*. Blackwell, London, 252 pp.

Delfanti, L., Colantoni, A., Recanatesi, F., Bencardino, M., Sateriano, A., Zambon, I., and Salvati, L. (2016). Solar plants, environmental degradation and local socioeconomic contexts: A case study in a Mediterranean country. *Environmental Impact Assessment Review*, 61, 88–93.

Di Feliciantonio, C. and Salvati, L. (2015). 'Southern' Alternatives of Urban Diffusion: Investigating Settlement Characteristics and Socio-Economic Patterns in Three Mediterranean Regions. Tijdschrift voor economische en sociale geografie, 106(4), 453–470.

Eakin, H., Lerner, A. M., and Murtinho, F. (2010). Adaptive capacity in evolving peri-urban spaces: responses to flood risk in the Upper Lerma River Valley, Mexico. *Global Environmental Change*, 20, 14–22.

Economidou, E. (1993). The Attic landscape throughout the centuries and its human degradation. *Landscape and Urban Planning*, 24 (1-4), 33–37.

Garcia Latorre, J., García-Latorre, J., and Sanchez-Picón, A. (2001). Dealing with aridity: socio-economic structures and environmental changes in an arid Mediterranean region. *Land Use Policy*, 18(1), 53–64.

Kairis, O., Karavitis, C., Salvati, L., Kounalaki, A., and Kosmas, K. (2015). Exploring the impact of overgrazing on soil erosion and land degradation in a dry Mediterranean agro-forest landscape (Crete, Greece). *Arid Land Research and Management* 29(3), 360–374.

Karamesouti, M., Detsis, V., Kounalaki, A., Vasiliou, P., Salvati, L., and Kosmas, C. (2015). Land-use and land degradation processes affecting soil resources: Evidence from a traditional Mediterranean cropland (Greece). *Catena*, 132, 45–55.

Kosmas, C., Karamesouti, M., Kounalaki, K., Detsis, V., Vassiliou, P., and Salvati, L. (2016). Land degradation and long-term changes in agro-pastoral systems: An empirical analysis of ecological resilience in Asteroussia-Crete (Greece). Catena, 147, 196–204.

Leontidou, L. (1990). The Mediterranean city in transition. *Cambridge University Press*, Cambridge, 296 pp.

Mairota, P., Thornes, J. B., and Geeson, N. (1998). Atlas of Mediterranean environments in Europe. The desertification context. Wiley, Chichester, 205 pp.

Paül, V., and Tonts, M. (2005). Containing urban sprawl: trends in land use and spatial planning in the metropolitan region of Barcelona. *Journal of Environmental Planning and Management*, 48(1), 7–35.

Recanatesi, F., Clemente, M., Grigoriadis, E., Ranalli, F., Zitti, M., and Salvati, L. (2016). A fifty-year sustainability assessment of Italian agro-forest districts. *Sustainability*, 8(1), 32.

Salvati, L. (2016). The Dark Side of the Crisis: Disparities in per Capita income (2000–12) and the Urban-Rural Gradient in Greece. *Tijdschrift voor economische en sociale geografie,* 107(5), 628–641.

Salvati, L., Ciommi, M. T., Serra, P., and Chelli, F. M. (2019). Exploring the spatial structure of housing prices under economic expansion and stagnation: The role of socio-demographic factors in metropolitan Rome, Italy. *Land Use Policy*, 81, 143–152.

Salvati, L., Gemmiti, R., and Perini, L. (2012). Land degradation in Mediterranean urban areas: an unexplored link with planning? Area, 44(3), 317–325.

Salvati, L., Sateriano, A., and Grigoriadis, E. (2016). Crisis and the city: profiling urban growth under economic expansion and stagnation. *Letters in Spatial and Resource Sciences*, 9(3), 329–342.

Salvati, L. and Zitti, M. (2007). Territorial disparities, natural resource distribution, and land degradation: a case study in southern Europe. *Geojournal*, 70(2), 185–194.

Salvati, L. and Zitti, M. (2009). The environmental 'risky' region: identifying land degradation processes through integration of socio-economic and ecological indicators in a multivariate regionalization model. *Environmental Management*, 44(5), 888.

Schneider, A. and Woodcock, C. E. (2008). Compact, dispersed, fragmented, extensive? A comparison of urban growth in twenty-five global cities using remotely sensed data, pattern metrics and census information. *Urban Studies*, 45(3), 659–692.

Smiraglia, D., Ceccarelli, T., Bajocco, S., Salvati, L., and Perini, L. (2016). Linking trajectories of land change, land degradation processes and ecosystem services. *Environmental Research*, 147, 590–600.

Weber, C., Petropoulou, C., and Hirsch, J. (2005). Urban development in the Athens metropolitan area using remote sensing data with supervised analysis and GIS. *International Journal of Remote Sensing*, 26(4), 785–796.

Zambon, I., Benedetti, A., Ferrara, C., and Salvati, L. (2018). Soil matters? A multivariate analysis of socioeconomic constraints to urban expansion in Mediterranean Europe. *Ecological Economics*, 146, 173–183.

Zambon, I., Colantoni, A., Carlucci, M., Morrow, N., Sateriano, A., and Salvati, L. (2017). Land quality, sustainable development and environmental degradation in agricultural districts: A computational approach based on entropy indexes. *Environmental Impact Assessment Review*, 64, 37–46.

Index

About the Editors

Antonio Tomao graduated in "Forestry and Environmental Sciences" from the University of Tuscia (Viterbo, Italy) in 2011 and holds a PhD in "Landscape and environment design management and planning" from the University of Rome "La Sapienza" in 2015. He is a fixed term researcher at the Council for Research in Agriculture and the Analysis of Agricultural Economics (CREA) and adjunct professor of "urban and territorial planning" at the Department of Economics and Law of the University of Macerata. Since 2011, he has been a collaborator in research activities and subsequently a research fellow (in role until 2021) at the Department for Innovation in Biological, Agro-food and Forest systems (DIBAF) of the University of Tuscia. He has carried out research periods abroad as a "visiting researcher" at the Forest Science and Technology Centre of Catalonia (CTFC) in Spain in 2015 and the Swedish University of Agricultural Sciences (SLU), in Sweden, in 2019. He has a background and expertise in geomatics applications to environmental monitoring, urban forest risk assessment, urban planning, regional planning and green infrastructure planning.

Matteo Clemente, Architect, PhD in Surveying and Representation of the Built Environment (1999). Research fellowship (2013/2017) at the Department of Architecture and Design of the University of Rome La Sapienza. He carries out research on the issues of redevelopment of the contemporary city, with particular reference to the re-design of public spaces, from the scale of urban design to that of landscape design and temporary set-up. He has an academic license for Associate Professor in Architectural Design and Urban Planning in 2018. He taught Exhibit and Public Design at the University of Rome La Sapienza, Architectural and Urban Design at the Federico II University of Naples, and Landscape Planning at the University of Tuscia. His publications include*: Re-design dello spazio pubblico, FrancoAngeli*, 2017, *Comporre e scomporre l'architettura, Aracne*, 2012, and *Estetica delle periferie urbane, Officina Edizioni, Roma*, 2005.

He is owner of a planning firm based in Rome mtstudio srl (www.mtstudio.it).